THE BIOSYNTHESIS OF DEOXYRIBOSE

C I B A LECTURES IN MICROBIAL BIOCHEMISTRY

1960 M. R. J. Salton, *Microbial Cell Walls*

1961 A. Kornberg, *Enzymatic Synthesis of DNA*

1962 B. L. Horecker, *Pentose Metabolism in Bacteria*

1963 J. W. Moulder, *The Psittacosis Group as Bacteria*

1967 P. Reichard, *The Biosynthesis of Deoxyribose*

Adirondack Community College Library
DISCARDED

CIBA LECTURES IN BIOCHEMISTRY

THE BIOSYNTHESIS OF DEOXYRIBOSE

BY PETER REICHARD

1967

John Wiley & Sons, Inc.

New York · London · Sydney · Toronto

The CIBA Lectures in Microbial Biochemistry were established in 1955 at the Institute of Microbiology, Rutgers, The State University of New Jersey, through the support of CIBA Pharmaceutical Products Inc., Summit, N. J. The lectures are given in the spring of each year at the Institute of Microbiology, New Brunswick, N. J.

18542

Copyright © 1968 by John Wiley & Sons, Inc.

All rights reserved

No part of this book may be reproduced by any means, nor transmitted, nor translated into a machine language without the written permission of the publisher.

Library of Congress Catalog Card Number: 68–8952
SBN 471 71496 8
Printed in the United States of America

PREFACE

The invitation to give the CIBA lectures at Rutgers University came at a time when our understanding of the enzymology in the synthesis of deoxyribose began to catch up with knowledge of the biosynthesis of other important carbohydrates. Progress in the field had been slow and the complexity of the enzyme system in deoxyribotide formation had been the cause of many frustrating experiences. In retrospect, however, this complexity was the source of much of the joy the problem gave and still gives to us who work with it. In many ways work on this specific problem turned out to be a collection of biochemical actualities which included a new hydrogen transport system involving the oxidoreduction of disulfide bonds, a new metabolic function of vitamin B_{12}, and the molecular mechanism of allosteric effects. Just now the work appears to lead into the realms of nonheme iron proteins and multienzyme systems.

I believe that these general biochemical aspects fully justify the publication of a book about an intrinsically quite limited subject. The material presented here in essence corresponds to my lectures at New Brunswick in May 1967 —with a few results obtained since that time.

I wish to thank the members of the Department of Microbiology of Rutgers University—and in particular Dr. J. O. Lampen—for their generous hospitality and many stimulating discussions during my stay at Rutgers University. I am also very grateful to CIBA for giving me the opportunity to come to New Brunswick.

Department of Chemistry 2
Karolinska Institutet, Stockholm P. REICHARD
March 1968

CONTENTS

THE BIOSYNTHESIS OF DEOXYRIBOSE

CHAPTER

1

SYNTHETIC PATHWAYS

The basic theme of this book is simple and may appear limited: it deals with the removal of a hydroxyl group and its replacement by hydrogen during the reduction of ribose to deoxyribose.

However, the simplicity of this reduction is deceptive. After more than 10 years of work on the problem, we realize that an intricate interplay between different enzymes and coenzymes is required for the formation of deoxyribonucleotides from ribonucleotides. The clarification of the mechanism of ribonucleotide reduction has resulted in the discovery of several phenomena that appear to be of general biochemical interest.

First, the work has led to the discovery of the thioredoxin system—a new type of hydrogen carrier system that involves the oxidoreduction of a specific disulfide bond of a small protein.

Second, it has included the finding that the coenzyme form of vitamin B_{12} participates as a hydrogen carrier in the ribonucleotide reductase from *Lactobacilli,* while in the enzyme system from *Escherichia coli,* and probably also in that from mammalian cells, a nonheme iron protein appears to substitute for cobamide coenzyme.

Finally, the investigations led to the discovery of the intriguing allosteric regulation of ribonucleotide reduction and provided a clear-cut example of a case in which the substrate specificity of an enzyme is determined by allosteric effectors.

I shall discuss these particular aspects in more detail in the two following chapters. First, I intend to provide a more general outline of the enzymatic synthesis of deoxyribose. At the end of this chapter, I will also discuss some of our recent studies on the inhibition of the enzyme reaction.

Biosynthesis of Deoxyribonucleotides

Function of Deoxyribose. In most living cells deoxyribose occurs almost exclusively bound in deoxyribonucleic acid. In 1961, Dr. A. Kornberg[1] in his Ciba lectures described work that led to the elucidation of the general mechanism for the enzymatic synthesis of DNA. Figure 1.1 shows that the net synthesis of DNA requires the polymerization of the four deoxyribonucleoside triphosphates by the enzyme DNA polymerase as directed by the template DNA.

The question of how deoxyribose is synthesized in the cell can thus be formulated in a more precise form: What is the mechanism for the biosynthesis of deoxyribonucleotides?

In principle, two main experimental approaches can be used to solve a problem of this kind:

1. One can try to define and study a series of enzyme reactions that lead to the formation of deoxyribonucleotides

$$\left.\begin{array}{l} n\,dTTP \\ n\,dGTP \\ n\,dCTP \\ n\,dATP \end{array}\right\} + DNA_{primer} \rightleftharpoons DNA - \left[\begin{array}{c} dTMP \\ dGMP \\ dCMP \\ dAMP \end{array}\right]_n + 4nPP$$

Fig. 1.1. The DNA polymerase reaction.[1]

from small molecules. This method involves the isolation and purification of enzymes.

2. Alternatively, one can administer the proper isotopically labeled compounds to intact animals or whole cell systems and study their utilization for the synthesis of deoxyribose of DNA.

Both of these approaches began around 1950.

Deoxyribose Aldolase Pathway. Figure 1.2 shows a sequence of four known enzyme reactions that could result in the formation of a deoxyribonucleotide. The key enzyme that catalyzes the first step of this sequence is deoxyribose aldolase. It was discovered by Racker[2,3] in 1951 in *E. coli* and was found subsequently to be distributed widely in different types of cells.[3–8] The enzyme catalyzes the reversible formation of deoxyribose 5-phosphate from acetaldehyde and glyceraldehyde phosphate. In the second step, deoxyribose 1-phosphate is formed by isomerization of the 5-phosphate,[9] and in the third step, a purine or pyrimidine is attached to deoxyribose by a nucleoside phosphorylase.[10–15] Finally, the deoxyribonucleoside thus formed is phosphorylated with ATP.[16] The last reaction is essentially irreversible and could pull the sequence of the first three freely reversible enzyme reactions in the direction of the synthesis of the deoxyribonucleotide.

There is, however, considerable evidence that excludes the reaction sequence outlined in Fig. 1.2 as a synthetic

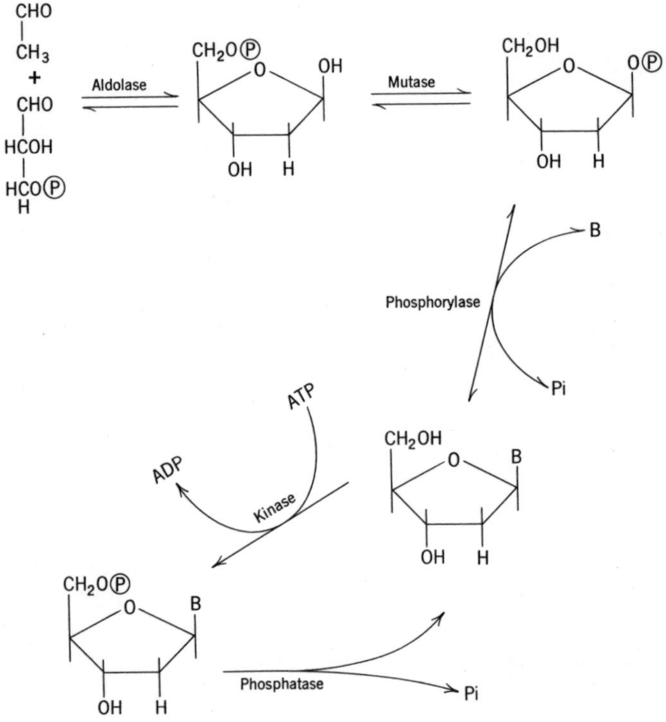

Fig. 1.2. The deoxyribose aldolase pathway.

pathway, and it seems more likely that the main function of the first three enzymes is the degradation of nucleotides. In the degradative pathway the dephosphorylation of the deoxyribonucleotide is catalyzed by a phosphatase. This does not imply that all the individual enzymes depicted in Fig. 1.2 exclusively have a catabolic function. Instead it seems clear that both phosphorylases and kinases participate in synthetic reactions involved in the reutilization of bases and nucleosides ("salvage" pathways) and possibly also in the transport of nucleic acid components across cell membranes.

There are at least four lines of evidence that speak against the thesis that the complete deoxyribose aldolase pathway of Fig. 1.2 participates in deoxyribose synthesis.

1. As Racker noted, deoxyribose aldolase has a rather low affinity for acetaldehyde. The concentration of acetaldehyde in cells is probably too low to allow an efficient synthesis of deoxyribose 5-phosphate. To circumvent this difficulty Racker suggested that an aldehyde linked to a purine or pyrimidine precursor may be the actual substrate for the enzyme. However, this type of compound has never been found.

2. Further negative evidence comes from isotope experiments carried out mainly in the laboratories of Cohen, Bernstein, Sable and Horecker, such as were recently reviewed by Sable.[17] Labeled small molecules (e.g., ^{14}C-glucose, ^{14}C-acetate, and $NaH\,^{14}CO_3$) were used as precursors for RNA and DNA in various types of cells. The incorporation of isotope into the different carbon atoms of ribose and deoxyribose was then measured and compared with the corresponding incorporation into glucose. Although the data were conflicting in some minor aspects, they clearly excluded in most cases the synthesis of deoxyribose via a condensation of a 2 carbon and a 3 carbon unit, as demanded by the deoxyribose aldolase pathway.

3. A third type of negative evidence derives from the substrate specificity of nucleoside phosphorylases. In the case of pyrimidines, only uracil or thymine are substrates; orotic acid is not utilized. On the other hand, orotic acid is the only pyrimidine synthesized *de novo;* no enzymes are known that catalyze the formation of either uracil or thymine. The

substrate specificity of nucleoside phosphorylases is therefore not suited for a *de novo* synthesis of deoxyribonucleosides.

4. A final and probably decisive argument arises from the occurrence of bacterial mutants that lack deoxyribose aldolase.[18] These mutants are not blocked in the synthesis of deoxyribose.

Ribonucleotide Reduction

***In vivo* Experiments.** The original evidence for another pathway of deoxyribose biosynthesis came from studies on the incorporation of ^{15}N-labeled pyrimidine nucleosides into the nucleic acids of the rat. These experiments, which were started in Stockholm before Racker's discovery of deoxyribose aldolase, are summarized in Fig. 1.3.

It was found that ^{15}N-labeled cytidine (and to a much smaller extent ^{15}N-uridine) is incorporated into both RNA and DNA by the rat.[19] As it was known from the work of Plentl and Schoenheimer[20] and Brown and his collabora-

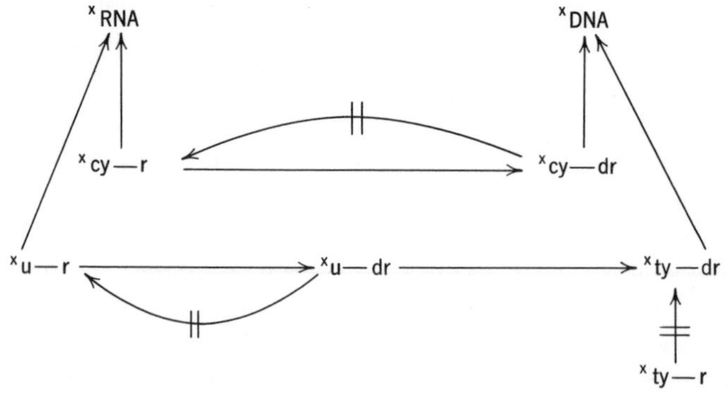

Fig. 1.3. Utilization of ^{15}N-nucleosides by the rat.

tors[21] that free pyrimidines (cytosine, uracil, and thymine) are not used for nucleic acid synthesis, it could be argued that the incorporation of the ribonucleosides had occurred without cleavage of the pyrimidine-sugar bond. Because labeled cytidine was recovered as labeled deoxycytidine in DNA, it was furthermore argued that the rat has the capacity to transform pyrimidine-bound ribose to pyrimidine-bound deoxyribose. This finding was the first indication of a pathway of deoxyribose biosynthesis involving the direct reduction of a ribosyl derivative.

Corroborative evidence for this concept came from experiments with labeled deoxyribonucleosides. Both ^{15}N-labeled deoxycytidine and deoxythymidine are incorporated into DNA, while no incorporation into RNA occurs.[22] Thus the postulated reduction of a ribosyl derivative must be an irreversible process (Fig. 1.3). It was then suggested that the exclusive incorporation of the deoxyribonucleosides into DNA might be of great importance in experiments that aim at a specific labeling of DNA, for example, during radioautography. The usefulness of this idea has now been well demonstrated.

Our experiments with ^{15}N-labeled nucleosides strongly suggested but did not prove the occurrence of a ribose reduction pathway. More decisive evidence came from experiments by Rose and Schweigert,[23] who used ^{14}C-cytidine labeled in both the sugar and the base as a precursor for DNA. They found that the ratio between the specific activities of cytosine and deoxyribose in the isolated DNA was the same as that between the specific activities of the pyrimidine and the sugar of the injected cytidine. This finding conclusively established that the ribose → deoxyribose transformation occurs without breakage of the pyrimidine-sugar bond.

Rose and Schweigert's experiments were published in

1953. Shortly thereafter several laboratories started a search for the appropriate enzyme system that, in the test tube, would catalyze the formation of a deoxyribonucleotide from a ribonucleotide. For a considerable period of time this search was unsuccessful and the first positive experiments did not appear until highly labeled radioactive ribonucleotides became available. The use of these materials as substrates made possible the demonstration of very low enzyme activities. Even then enzyme activity was found only in cells with a very high rate of DNA synthesis, i.e., certain bacteria, tumor cells, embryonic cells, and tissue culture cells. In all cases only low enzyme activities could be demonstrated in crude cell extracts. It is therefore not surprising that it was proposed on several occasions[5,24] that the reduction of ribonucleotides may represent only a minor pathway for the synthesis of deoxyribose—active only under special conditions. Accordingly it was argued that other major pathways of deoxyribose synthesis exist such as the one illustrated by Fig. 1.2.

I should like to illustrate the fallacies of such an argument with a recent example that concerns regenerating rat liver. Between 24 and 48 hr after partial hepatectomy a rapid and nearly synchronous synthesis of DNA occurs in the remaining liver. This obviously requires a considerable supply of deoxyribonucleotides. In fact, one can estimate that liver cells on this occasion synthesize deoxyribose at about the same rate as do embryonic cells or tumor cells. Nevertheless, attempts to find an active ribonucleotide reductase in extracts from regenerating liver were negative. On the other hand, the activity of deoxyribose aldolase, which is fairly high in normal rat liver, appears to increase during liver regeneration, and it thus appears reasonable to postulate that this enzyme plays an important role in deoxyribose synthesis in liver.

Larsson and Neilands[25] recently evaluated the relative contributions of the two possible pathways of deoxyribose synthesis in regenerating rat liver. Hepatectomized rats were simultaneously injected with ^{14}C-cytidine and $NaH_2{}^{32}PO_4$ and the incorporation of isotope into the nucleotides of liver RNA and DNA was measured.

Figure 1.4 gives the concept of the experiment. The injection of isotopic inorganic phosphate results in a rapid labeling of the terminal phosphate group of ATP used for the phosphorylation of the injected ^{14}C-cytidine. Any dCMP formed by the reduction of the CMP should have the same relative specific activity (i.e., the same ^{32}P:^{14}C ratio) as the CMP from which it was derived. On the other hand, a large excess of ^{32}P would be found in dCMP if it were formed by phosphorylation of nonlabeled deoxycytidine (synthesized

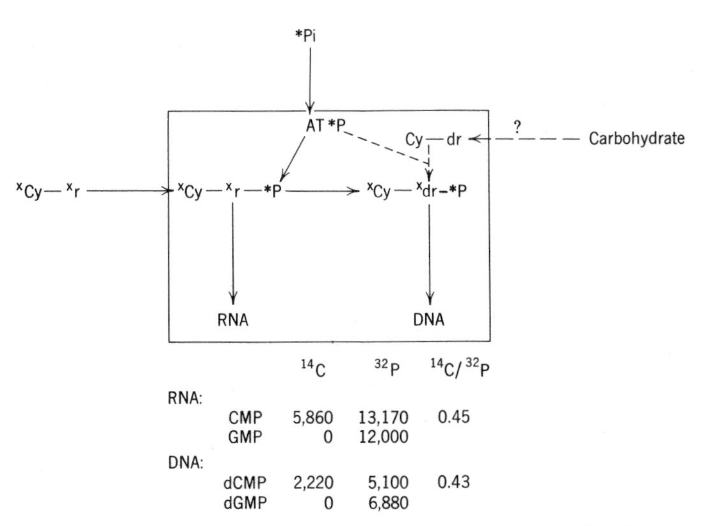

		^{14}C	^{32}P	^{14}C/^{32}P
RNA:				
	CMP	5,860	13,170	0.45
	GMP	0	12,000	
DNA:				
	dCMP	2,220	5,100	0.43
	dGMP	0	6,880	

Fig. 1.4. Comparison of utilization of $^{32}P_i$ and ^{14}C-cytidine by the rat.

via the aldolase pathway). A comparison of the ^{32}P:^{14}C ratios in CMP (from RNA) and dCMP (from DNA) gives the relative contributions of the two hypothetical pathways.

In the actual experiment an identical ratio was observed in CMP and dCMP, indicating that the contribution of the deoxyribose aldolase pathway was negligible and that all deoxyribose had been synthesized by the reduction of a ribosyl derivative.

These results were obtained after many failures to find an active ribonucleotide reductase in extracts from regenerating rat liver. They provided the necessary impetus for a renewed attack on the problem. After a systematic study, the experimental conditions were finally found that allowed the demonstration of the enzyme in the extract and it was possible to start with the purification of the enzyme system.[26] The partially purified liver enzyme appeared in all respects to be similar to the enzyme preparation from Novikoff hepatoma described earlier by Moore and Hurlbert.[27–29] Adult rat liver contains only minimal amounts of ribonucleotide reductase, but approximately 24 hr after hepatectomy a large increase in enzyme activity is observed (Fig. 1.5).

I have discussed the results with regenerating liver at some length because they serve to illustrate some of the difficulties that must be overcome before an active ribonucleotide reductase can be demonstrated in crude tissue extracts. In this particular instance the maintenance of the proper level of substrate phosphorylation and therefore the choice of the correct concentrations of ATP and Mg^{2+} appear to present the main problem. A further important problem is the maintenance of reducing power (TPNH or dithiol) during the course of the incubation. Finally, crude extracts usually contain inhibitors that may offer additional complications.

Ribonucleotide reductase from E. coli. Extracts from *E.*

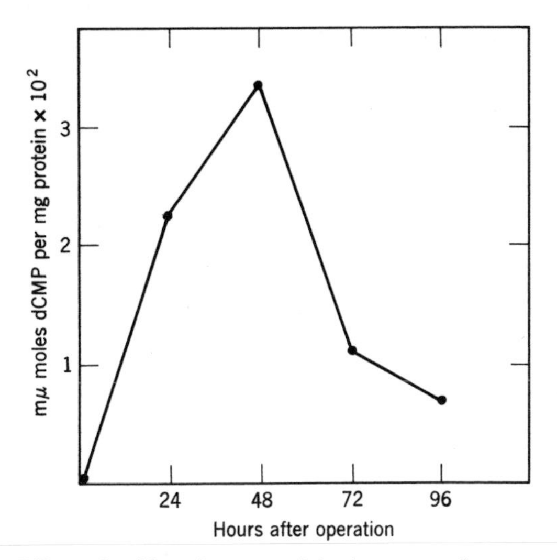

Fig. 1.5. Ribonucleotide reductase activity in extracts from regenerating rat liver at different times after hepatectomy.[26]

coli B provided the first reproducible source of enzymes catalyzing the reduction of ribonucleotides.[30,31] Optimal formation of deoxycytidine phosphates from CMP required addition of TPNH, ATP, and Mg^{2+}. The "enzyme reaction" could be represented schematically in the following way:

$$\text{CMP} \xrightarrow[\text{ATP, Mg}^{2+}]{\text{TPNH}} \text{deoxycytidine phosphates.} \tag{1.1}$$

TPNH was assumed to function as hydrogen donor in the process. The requirement for ATP and Mg^{2+}, however, was more difficult to understand and remained an enigma for a considerable time. Part of the explanation came when it was found that CDP—not the monophosphate—is the actual substrate in the enzyme reaction,[31] and that ATP is thus required for the phosphorylation of CMP.

The chemistry involved in the reduction of a ribonucleotide is shown in Fig. 1.6. The OH-group at position 2′ of a ribonucleoside diphosphate is replaced by a hydrogen. In Chapter 3 it will be shown that the same enzyme system catalyzes the reduction of all four commonly occurring ribonucleoside diphosphates and that the base therefore can be either cytosine, uracil, adenine, or guanine. For technical reasons most of our work was carried out with CDP as substrate and the complex enzyme system carrying out the formation of dCDP from CDP was originally called the CDP reductase system. However, a more general—and more proper—name for the enzyme system is "ribonucleoside diphosphate reductase system."

The complexity of the system made initial efforts to purify the enzyme a difficult task. A turning point came when reduced lipoate was used as hydrogen donor in place of TPNH.[32] It was then possible to separate and purify two protein fractions, provisionally called Fraction A and Fraction B, both required for the synthesis of deoxycytidine phosphates from CMP. The latter fraction could be further separated into Fractions B1 and B2.

Highly purified Fraction A was found to be identical with CMP kinase. It was required only when CMP was used as substrate. Fractions B1 and B2 were components of the

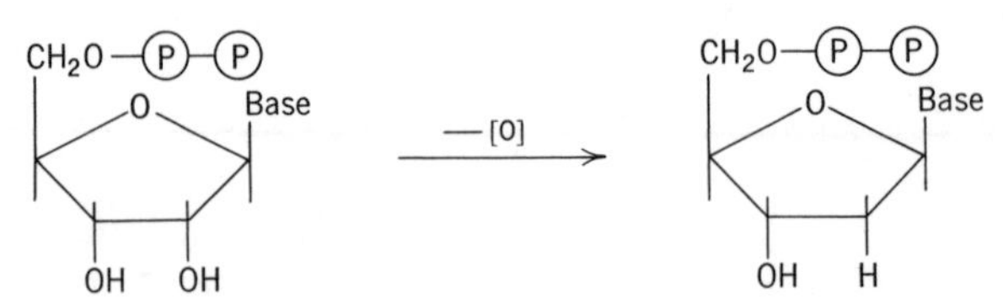

Fig. 1.6. Chemistry of ribonucleotide reduction in *E. coli*.

actual ribonucleoside diphosphate system. The results are summarized in the following scheme:

$$\text{CMP} \xrightarrow[\text{ATP, Mg}^{2+}]{\text{Fraction A}} \text{CDP} + \text{lipoate-(SH)}_2 \xrightarrow[\text{Mg}^{2+}\text{, ATP}]{\text{Fractions B1 + B2}} \text{dCDP} + \text{lipoate-S}_2. \quad (1.2)$$

Thus with the purified Fractions B1 and B2, reduced lipoate functions as hydrogen donor. The requirements for ATP and Mg^{2+} were also evident with CDP as substrate.

It appeared improbable that reduced lipoate was the physiological hydrogen donor in the reaction. It seemed more likely that enzyme purification removed a substance, probably a dithiol (or a disulfide, reducible with TPNH) that could be replaced by reduced lipoate.

The presence of such a substance could indeed be demonstrated in Fraction A and after extensive purification a protein was obtained in essentially pure form that, together with TPNH, substituted for reduced lipoate when the reduction of CDP was catalyzed by Fraction B. This protein was named thioredoxin.[33] It contained a single disulfide bridge (thioredoxin-S_2) that was reduced to a dithiol (thioredoxin-$(SH)_2$) by a specific enzyme, called thioredoxin reductase,[34] present as an impurity in Fraction B:

$$\text{Thioredoxin-S}_2 + \text{TPNH} + \text{H}^+ \underset{\text{reductase}}{\overset{\text{thioredoxin}}{\rightleftharpoons}} \text{thioredoxin-(SH)}_2 + \text{TPN}^+. \quad (1.3)$$

Thioredoxin reductase was purified from *E. coli* and found to be a flavoprotein.[34] The enzyme, together with thioredoxin and TPNH, forms a hydrogen transport system that functions in the reduction of ribonucleotides. I will discuss the components of this system in detail in the next chapter.

The reduction of a ribonucleotide to the corresponding deoxyribonucleotide by proteins B1 + B2 can thus occur with either reduced lipoate or reduced thioredoxin. The latter compound is the physiological hydrogen donor, while reduced lipoate acts as a model compound. With thioredoxin as the hydrogen donor, the reduction of the ribonucleotide requires either substrate amounts of thioredoxin-$(SH)_2$ or catalytic amounts of thioredoxin-S_2 and substrate amounts of TPNH. In the latter case thioredoxin reductase is required for the continuous regeneration of reduced thioredoxin-$(SH)_2$. These interrelationships are illustrated in Fig. 1.7.

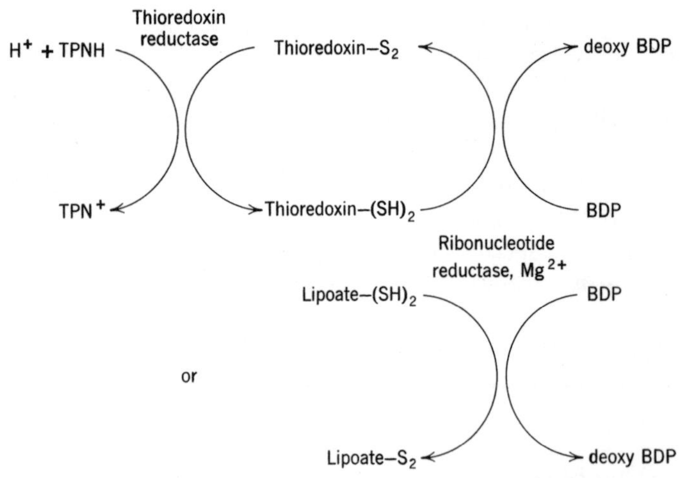

Fig. 1.7. Reduction of ribonucleotides by thioredoxin system or reduced lipoate.

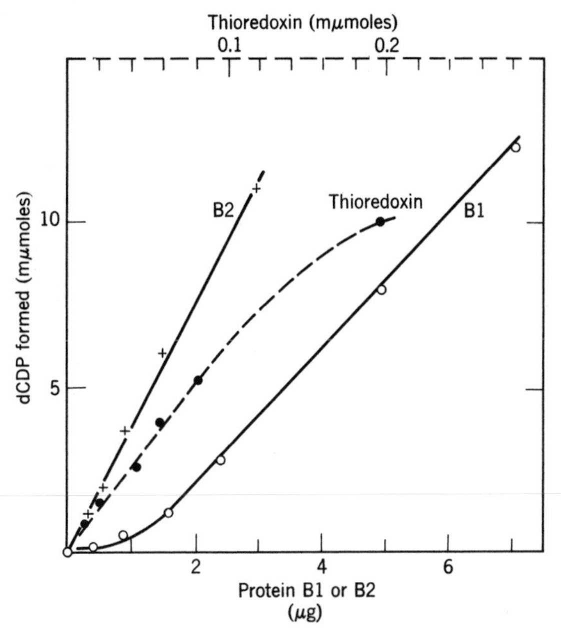

Fig. 1.8. Assays of protein B1, protein B2, or thioredoxin.

With catalytic amounts of thioredoxin, one mole of TPNH is oxidized during the formation of each mole of dCDP. This forms the basis of a rapid assay method [35] for the different components of the CDP reductase system, as the disappearance of TPNH can be followed by measuring the decrease in the absorbance at 340 mμ. Such assays are illustrated for protein B1, protein B2, and thioredoxin in Fig. 1.8. In the presence of an excess of two of the components, the reaction rate is more or less proportional to the amount of the third component added. The method has its greatest merit in the assay of proteins B1 and B2, while thioredoxin (and thioredoxin reductase) usually can be assayed directly according to reaction (1.3).

With the aid of these assay methods it was possible to purify proteins B1 and B2, thioredoxin reductase, and thioredoxin from *E. coli*.[33–36] A general outline of the purification procedures for the four proteins is given in Fig. 1.9. Both thioredoxin and thioredoxin reductase were obtained in pure form. Because of the difficulties discussed below we can not be absolutely certain of the purity of our preparations of proteins B1 and B2, but have now good indications that both are essentially pure.

Together, proteins B1 and B2 constitute the enzyme ribonucleoside diphosphate reductase. In Chapter 3, I will pre-

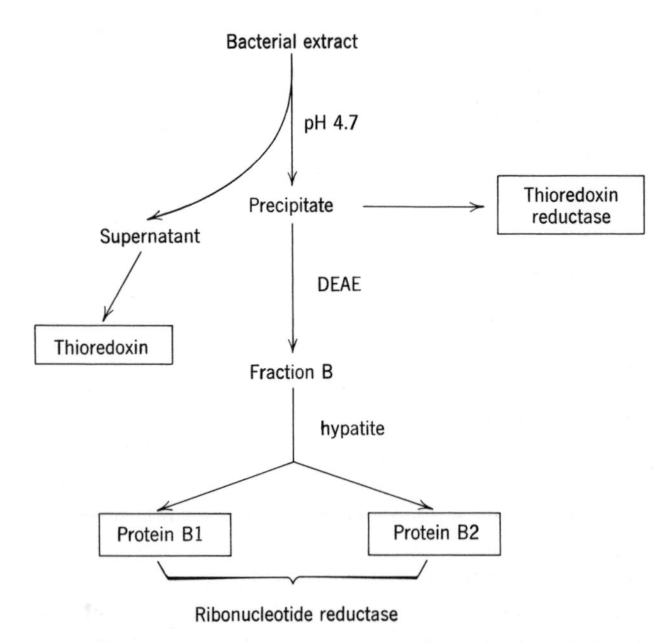

Fig. 1.9. Separation of components of ribonucleoside diphosphate reductase system from *E. coli*.

sent evidence showing that the two proteins are nonidentical subunits that in the presence of Mg^{2+} form the active enzyme. Now I should like to summarize briefly some of the properties of the isolated proteins.

Both proteins presented considerable difficulties during purification and, unless special precautions were taken, severe losses of enzyme activity occurred. Protein B1 required the presence of 0.01 M dithioerythrol for maximal stability. On ultracentrifugation the best preparations of protein B1 migrate as an essentially homogenous peak with an $s_{20,w}$ value of about 7.8 S. A small impurity (with $s_{20,w}$ of 5.2 S) is always observed. On storage at room temperature inactivation of the enzyme occurred, paralleled by a transfer of material from the 7.8 S peak to the 5.2 S peak.[36]

Another type of difficulty was encountered with protein B2. In this case the recovery of enzyme activity at later stages of purification was highly irreproducible and in certain cases, for example, on prolonged ultrafiltration against buffers containing EDTA, complete loss of enzyme activity could occur accompanied by precipitation of the protein. The explanation of these difficulties came when it was found that protein B2 contained iron and that the presence of the bound metal was required for the catalytic function of the protein.[37]

Several lines of evidence lead to this conclusion. Thus enzyme activity and iron content were enriched in parallel during purification. The purest preparations of protein B2 contain close to 2 moles of iron per mole of enzyme. This iron is bound very tightly, but it can be removed on prolonged dialysis against 8-hydroxyquinoline. The resulting apoprotein is inactive and can be reactivated completely by the addition of ionic iron.

Protein B2 has a very characteristic spectrum with a quite sharp peak around 410 mμ and a second broader maximum around 360 mμ (Fig. 1.10). Both peaks are lost

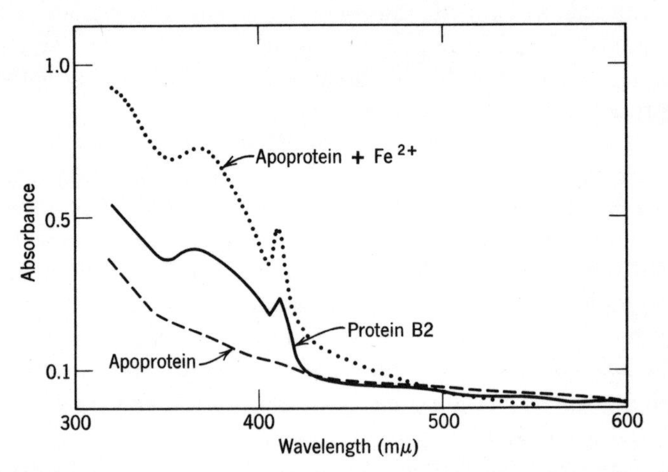

Fig. 1.10. Spectra of protein B2, iron-free protein B2, and reactivated protein B2.[37]

when iron is removed from the protein and reappear when the metal is reintroduced. There appears to be a clear connection between the spectrum of protein B2, its iron content, and its enzymatic activity.

Nothing is known about the mode of binding of the iron within the protein. Unlike most other nonheme iron proteins,[38] protein B2 contains no "inorganic sulfide." The very sharp peak at 410 mμ also differentiates protein B2 from other nonheme iron proteins. I will return to the possible function of iron in ribonucleotide reductase in connection with a discussion of the mechanism of the reaction.

Other Ribonucleotide Reductases. Two other ribonucleotide reductase systems have been studied in considerable detail.

An enzyme system from Novikoff hepatoma was partially purified by Moore and collaborators.[27–29] Until recently this system was the only mammalian ribonucleotide reductase that had been investigated closely. In most respects the mammalian system shows a close similarity to the reductase from *E. coli.* It appears to consist of two subunits[39] and to utilize ribonucleoside *di*phosphates as substrates and a thioredoxin system as hydrogen transport system.[40] Furthermore, the enzyme is subject to an allosteric regulatory mechanism[29] quite similar to that found for the *E. coli* reductase. This point will be discussed in Chapter 3. Finally, the enzyme is strongly stimulated by the addition of $FeCl_3$,[27] suggesting the participation of iron in the reaction mechanism.

A somewhat different type of ribonucleoside reductase was discovered by Blakley and Barker[41] in *Lactobacillus leichmannii* and later studied in detail in the laboratories of Abrams[42,43], Beck,[44,45] and Blakley.[46,47] The most conspicuous difference concerns an absolute requirement of the enzyme for cobamide coenzyme. Prior to isolation of this enzyme system, nutritional experiments[48] and isotope incorporation data[49] had already suggested the involvement of a vitamin B_{12} derivative in deoxyribose synthesis in *L. leichmannii.* After many negative efforts in several different laboratories, including our own, Blakley and Barker in 1964 successfully prepared cell-free extracts from *L. leichmannii* that catalyzed the reduction of CMP to dCMP.[41] The reaction was strongly dependent on the addition of dimethylbenzimidazolyl cobamide coenzyme and, furthermore, was stimulated by ATP, Mg^{2+}, and a TPNH regenerating system.

Further work with purified enzymes delineated the similarities and dissimilarities between the ribonucleotide reductases from *E. coli* and *L. leichmannii.* The latter enzyme does not appear to contain nonidentical subunits; it shows no absolute requirement for Mg^{2+} and no requirement for iron. Ribonucleoside *tri*phosphates serve as substrates for the enzyme from *L. leichmannii,* and the reaction shows an absolute requirement for cobamide coenzyme.

The similarities between the two microbial enzymes include the ability to use dithiols as hydrogen donors, the function of a thioredoxin system as the physiological electron transport system, and the sensitivity to certain allosteric effects that influence the substrate specificity of the enzyme.

A summary of the properties of the three ribonucleotide reductase systems is given in Table 1.1.

TABLE 1.1.

Comparison of Properties of Different Ribonucleotide Reductases

	E. coli	Novikoff hepatoma	*L. leichmannii*
Level of phosphorylation of substrate	Diphosphate	Diphosphate	Triphosphate
Requirement for cobamide coenzyme	No	No (?)	Yes
Hydrogen donor	Thioredoxin	Thioredoxin	Thioredoxin
General inhibition by dATP	Yes	Yes	No
Separation into nonidentical subunits	Yes	Yes	No
Requirement for Mg^{2+}	Absolute	Absolute	Relative
Involvement of iron	Yes	Yes	No

Inhibitors of Ribonucleotide Reduction

The reduction of ribonucleotides is a key step in the biosynthesis of DNA. Inhibition of this reaction is therefore of interest whenever one wishes to inhibit DNA synthesis. Earlier *in vivo* experiments had indicated that hydroxyurea, [50–52] cytosine arabinoside diphosphate,[53] and naladixic acid[54] were potential inhibitors of this step. In preliminary experiments[55] with the highly purified and reconstructed ribonucleotide reductase system from *E. coli,* we found that all three compounds inhibited the transformation of CDP to dCDP at concentrations of 10^{-3} to 10^{-2} M.

We then investigated in more detail the effects of one of these compounds and attempted to localize the site of inhibition; hydroxyurea was chosen for this purpose.[56]

Hydroxyurea. The first clue to the mechanism of action of this drug was obtained when it was found that the degree of inhibition was dependent on the relative proportions of protein B1 and protein B2 in the incubation mixture. The presence of a large excess of protein B2 resulted in a low degree of inhibition. A second important clue was the finding that the effect of hydroxyurea was

TABLE 1.2.

Effect of Preincubation with Hydroxyurea on Components of Ribonucleotide Reductase from *E. coli*

Preincubated component	Time of preincubation* (min)	dCDP formed (mμmoles)	Remaining activity (%)
Protein B1 + B2	0	2.70	100
	1	2.72	100
	2	2.23	82
	5	1.39	51
	10	0.43	16
Protein B1	0	5.60	100
	30	4.20	75
Protein B2	0	11.50	100
	30	0.10	1

* Preincubation at 23° with 10^{-2} M hydroxyurea.

time-dependent. When the enzyme was preincubated with hydroxyurea for different periods of time before the addition of substrate, the degree of inhibition was dependent on the length of the preincubation period (Table 1.2). This effect was limited to protein B2, and preincubation of this

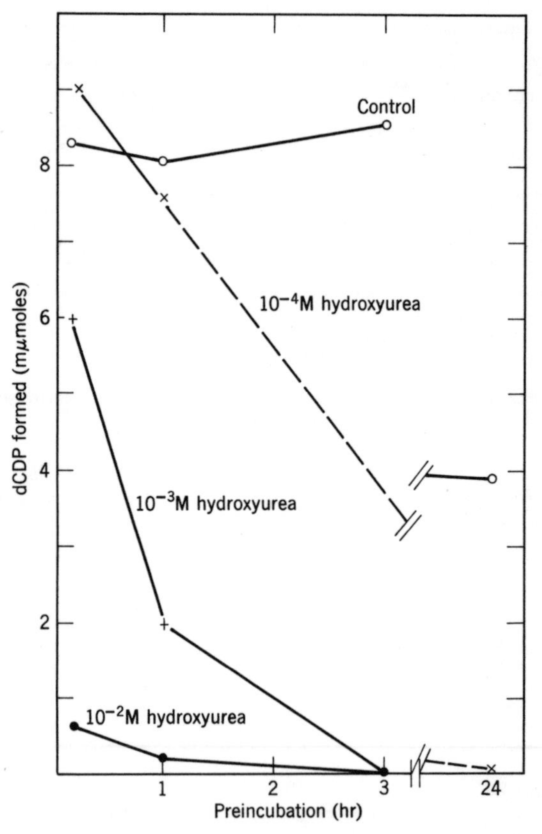

Fig. 1.11 Effect of preincubation with different concentrations of hydroxyurea on activity of protein B2.[56]

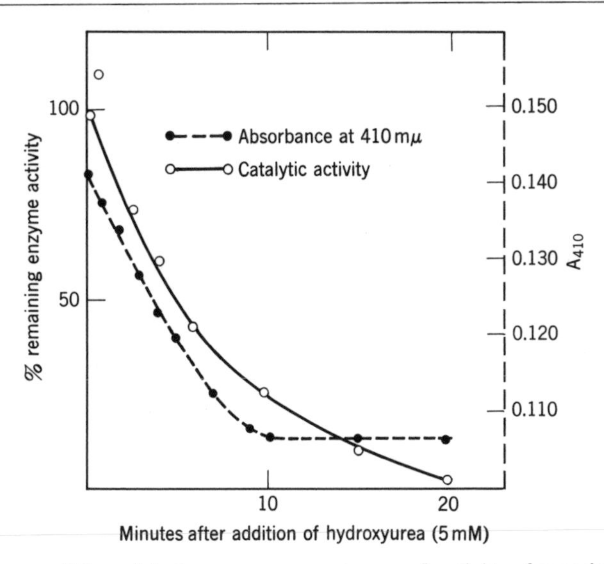

Fig. 1.12. Effect of hydroxyurea on spectrum and activity of protein B2.

protein alone with hydroxyurea gave complete inhibition, while preincubation of protein B1 was without effect.

A more detailed study of the effects of preincubation of protein B2 with different concentrations of hydroxyurea is described in Fig. 1.11. It is evident that rather low concentrations of hydroxyurea (10^{-4} M) could effect the complete inactivation of protein B2 if incubation was made for a prolonged period of time.

As mentioned earlier, protein B2 has a characteristic spectrum with a sharp peak around 410 mμ. Incubation of a solution of protein B2 with hydroxyurea results in a time-dependent disappearance of this peak.[57] The decrease in absorbance at 410 mμ parallels the loss of enzyme activity closely (Fig. 1.12). Similar results were obtained with hydroxylamine or hydrazine in place of hydroxyurea.

I have earlier discussed a connection between the peak at 410 mμ and the presence of iron in protein B2. The results with hydroxyurea demonstrate that the drug specifically inactivates protein B2 and also suggest that the inactivation is caused by some modification in the binding of iron in the protein.

It seems likely that the inhibitory effect of hydroxyurea on DNA synthesis *in vivo* is explained by the specific inhibition of protein B2 of the ribonucleotide reductase.

REFERENCES

1. Kornberg, A., *Enzymatic Synthesis of DNA,* Ciba Lectures 1961, Wiley, New York and London.
2. Racker, E., *Nature,* **167,** 408 (1951).
3. Racker, E., *J. Biol. Chem.,* **146,** 347 (1952).
4. McGeown, M. G., and F. H. Malpress, *Nature,* **170,** 575 (1952).
5. Boxer, G. E., and C. E. Shonk, *J. Biol. Chem.,* **233,** 535 (1958).
6. Roscoe, H. G., and W. L. Nelson, *J. Biol. Chem.,* **239,** 8 (1964).
7. Pricer, W. E., Jr., and B. L. Horecker, *J. Biol. Chem.,* **235,** 1292 (1960).
8. Jiang, N. S., and D. P. Groth, *J. Biol. Chem.,* **237,** 3339 (1962).
9. Hoffmann, C. E., and J. O. Lampen, *J. Biol. Chem.,* **198,** 885 (1952).
10. Manson, L. A., and J. O. Lampen, *J. Biol. Chem.,* **191,** 95 (1951).
11. Klein, W., *Z. Physiol. Chem.,* **231,** 125 (1935).
12. Kalckar, H. M., *J. Biol. Chem.,* **158,** 723 (1945).
13. Friedkin, M., and H. M. Kalckar, *J. Biol. Chem.,* **184,** 437 (1950).
14. Friedkin, M., *J. Biol. Chem.,* **184,** 449 (1950).
15. Friedkin, M., and D. Roberts, *J. Biol. Chem.,* **207,** 245 (1954).
16. Bollum, F. J. and V. R. Potter, *J. Biol. Chem.,* **233,** 478 (1958).
17. Sable, H. Z., *Advan. Enzymol.,* **28,** 391 (1966).
18. Breitman, T. R., and R. M. Bradford, *Biochim. Biophys. Acta,* **138,** 217 (1967).
19. Hammarsten, E., P. Reichard, and E. Saluste, *J. Biol. Chem.,* **183,** 105 (1950).
20. Plentl, A. A., and R. Schoenheimer, *J. Biol. Chem.,* **153,** 203 (1944).
21. Bendich, A., H. Getler, and G. B. Brown, *J. Biol. Chem.,* **177,** 565 (1949).

22. Reichard, P., and B. Estborn, *J. Biol. Chem.*, **188,** 839 (1951).
23. Rose, I. A., and B. S. Schweigert, *J. Biol. Chem.*, **202,** 635 (1953).
24. Groth, D. P., and N. Jiang, *Biochem. Biophys. Res. Commun.*, **22,** 62 (1966).
25. Larsson, A., and J. B. Neilands, *Biochem. Biophys. Res. Commun.*, **25,** 222 (1966).
26. Larsson, A., unpublished results.
27. Moore, E. C., and R. B. Hurlbert, *Biochim. Biophys. Acta*, **55,** 651 (1962).
28. Moore, E. C., and P. Reichard, *J. Biol. Chem.*, **239,** 3453 (1964).
29. Moore, E. C., and R. B. Hurlbert, *J. Biol. Chem.*, **241,** 4802 (1966).
30. Reichard, P., and L. Rutberg, *Biochim. Biophys. Acta*, **37,** 554 (1960).
31. Reichard, P., A. Baldesten, and L. Rutberg, *J. Biol. Chem.*, **236,** 1150 (1961).
32. Reichard, P., *J. Biol. Chem.*, **237,** 3513 (1962).
33. Laurent, T. C., E. C. Moore, and P. Reichard, *J. Biol. Chem.*, **239,** 3436 (1964).
34. Moore, E. C., P. Reichard, and L. Thelander, *J. Biol. Chem.*, **239,** 3445 (1964).
35. Holmgren, A., P. Reichard, and L. Thelander, *Proc. Natl. Acad. Sci. U.S.*, **54,** 830 (1965).
36. Brown, N. C., A. Larsson, P. Reichard, and L. Thelander, in preparation.
37. Brown, N. C., R. Eliasson, P. Reichard, and L. Thelander, *Biochem. Biophys. Res. Commun.*, **30,** 522 (1968).
38. Malkin, R., and J. C. Rabinowitz, *Ann. Rev. Biochem.*, **36,** 113 (1967).
39. Moore, E. C., personal communication.
40. Moore, E. C., *Biochem. Biophys. Res. Commun.*, **29,** 264 (1967).
41. Blakley, R. L., and H. A. Barker, *Biochem. Biophys. Res. Commun.*, **16,** 391 (1964).
42. Abrams, R., and S. Duraiswami, *Biochem. Biophys. Res. Commun.*, **18,** 409 (1965).
43. Abrams, R., *J. Biol. Chem.*, **240,** PC3697 (1965).
44. Beck, W. S., and J. Hardy, *Proc. Natl. Acad. Sci. U.S.*, **54,** 286 (1965).
45. Goulian, M., and W. S. Beck, *J. Biol. Chem.*, **241,** 4233 (1966).
46. Vitols, E., and R. L. Blakley, *Biochem. Biophys. Res. Commun.*, **21,** 466 (1965).

47. Ghambeer, R. K., and R. L. Blakley, *J. Biol. Chem.,* **241,** 4710 (1966).
48. Snell, E. E., E. Kitay, and W. S. McNutt, *J. Biol. Chem.,* **175,** 473 (1948).
49. Downing, M., and B. S. Schweigert, *J. Biol. Chem.,* **220,** 521 (1956).
50. Frenkel, E. P., W. N. Skinner, and J. D. Smiley, *Cancer Chemother. Rep.* **40,** 19 (1964).
51. Adams, R. L. P., R. Abrams, and I. Lieberman, *J. Biol. Chem.,* **241,** 903 (1966).
52. Neuhard, J., *Biochim. Biophys. Acta,* **145,** 1 (1967).
53. Chu, M. Y., and G. A. Fischer, *Biochem. Pharmacol.,* **11,** 423 (1962).
54. Goss, W. A., W. H. Deitz, and T. M. Cook, *J. Bacteriol.,* **89,** 1068 (1965).
55. Szabo, L., and P. Reichard, unpublished observations.
56. Krakoff, I., N. C. Brown, and P. Reichard, *Cancer Research,* in press.
57. Brown, N. C., and P. Reichard, unpublished observations.

CHAPTER

2

THE THIOREDOXIN SYSTEM

Hydrogen Transport. When we first attempted to purify ribonucleotide reductase from *E. coli,* our efforts were frustrated by the apparent unusual instability of the enzyme. Each purification step resulted in the loss of a large part of enzyme activity.

These losses did not primarily originate from denaturation of a delicate protein structure during purification. Instead, most of our difficulties arose from the fact that the reduction of a nucleotide such as CMP requires the cooperation of five different protein fractions and that all our purification procedures resulted in the separation of the components of a multienzyme system.

As discussed in the previous chapter, enzyme purification became feasible when we used reduced lipoate as hydrogen donor in place of TPNH and thereby obviated the requirement for two of the five protein fractions.[1] After considerable purification of the actual ribonucleoside diphosphate reductase (proteins B1 and B2) we could then reintroduce

TPNH as hydrogen donor and start the purification of the two remaining protein fractions. In this way thioredoxin[2] and thioredoxin reductase[3] were discovered, purified independently and eventually obtained in pure form. Together with TPNH the two proteins form an electron transport system that was named the *thioredoxin system*. It makes the hydrogen of TPNH available for the reduction of the secondary alcohol of the ribosyl moiety of the ribonucleotide.

First, I shall discuss briefly the methods for the assay of thioredoxin and thioredoxin reductase. The thioredoxin system can be coupled to the reduction of ribonucleotides (see Fig. 1.7), and in the presence of an excess of ribonucleotide reductase the formation of radioactive dCDP can be limited either by the amount of thioredoxin or thioredoxin reductase. This type of assay has the advantage that it can be used with very crude protein fractions; however, its complexity represents a distinct disadvantage. Accordingly, after some purification, it is more convenient to use a direct assay that involves only the reaction catalyzed by thioredoxin reductase:

$$\text{TPNH} + \text{H}^+ + \text{thioredoxin-S}_2 \rightleftharpoons \text{TPN}^+ + \text{thioredoxin-(SH)}_2. \qquad (2.1)$$

With substrate amounts of oxidized thioredoxin, one can either use the appearance of SH-groups or the disappearance of TPNH as a measure of the reaction. An example of the latter method is given in Fig. 2.1. The three curves show the decrease in absorbance at 340 mμ in experiments in which a constant amount of thioredoxin and increasing amounts of thioredoxin reductase are used. The initial slopes of each curve are proportional to the amounts of thioredoxin reductase added to the system, while the final plateau—which is identical for all three curves—is a measure of the amount of thioredoxin. At neutral pH the

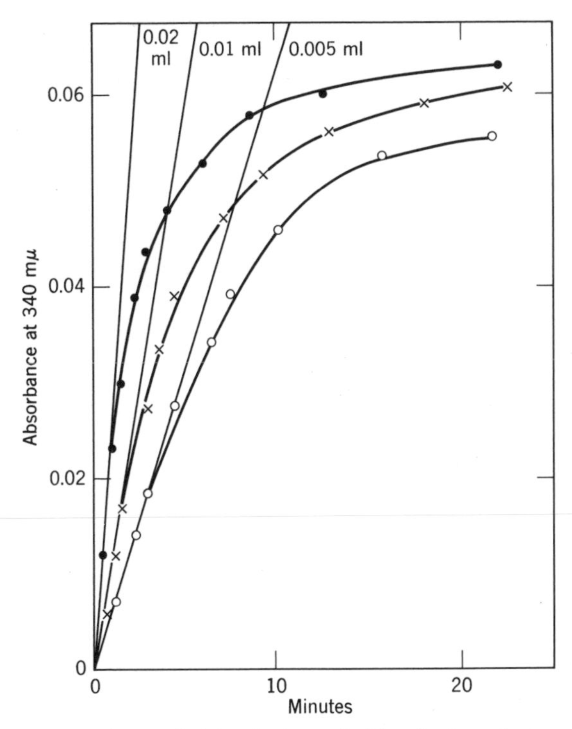

Fig. 2.1. Assay of thioredoxin and thioredoxin reductase.

equilibrium of reaction (2.1) is almost completely shifted towards the reduction of thioredoxin.

Thioredoxin Reductase. This enzyme was obtained in a homogenous form by Thelander.[4,5] It has a molecular weight of 66,000 and consists of two subunits that are probably identical. Each subunit contains one molecule of FAD, but no metal. Solutions of pure thioredoxin reductase have an intense yellow color and give the spectrum shown in Fig. 2.2. For comparison the spectrum of FAD

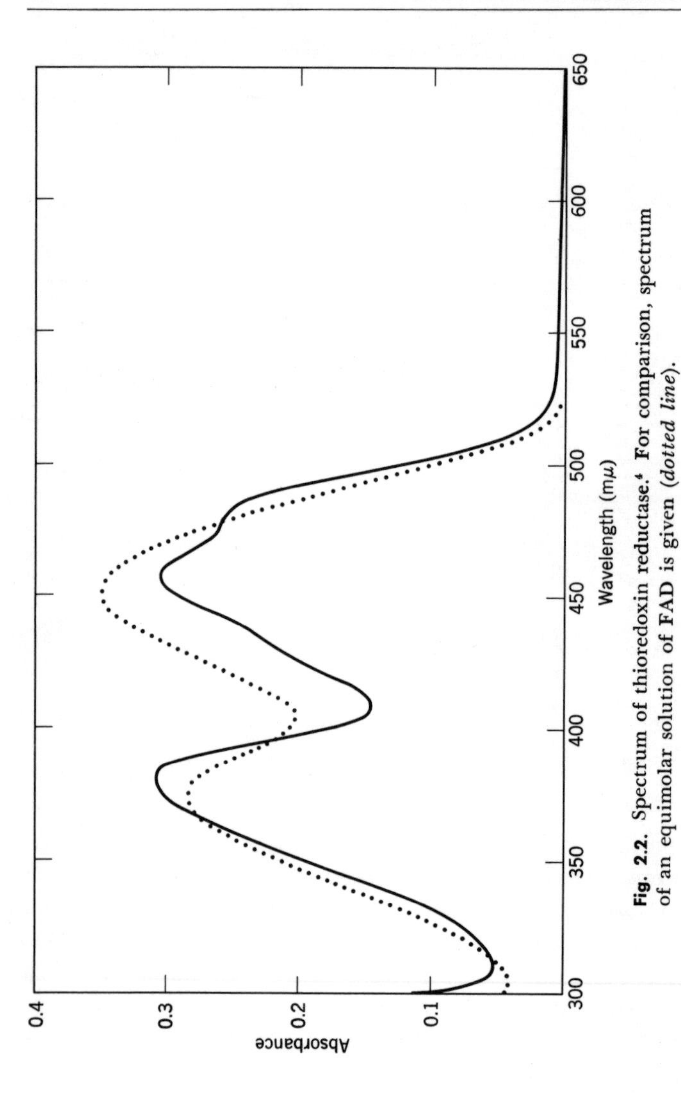

Fig. 2.2. Spectrum of thioredoxin reductase.[4] For comparison, spectrum of an equimolar solution of FAD is given (*dotted line*).

is also shown in the figure. Each subunit of the enzyme contains four half-cystine residues; two of these occur in the reduced form, while the other two form a disulfide bridge. The disulfide is reduced by TPNH and appears to participate in the catalytic mechanism. In this and in many other respects, thioredoxin reductase resembles two other flavoproteins—glutathion reductase and dihydrolipoic dehydrogenase—that also catalyze the reductions of S_2-bonds by reduced pyridine nucleotides. The mechanism of catalysis for these two flavoproteins was extensively investigated earlier by Massey and coworkers.[6–8]

Thioredoxin reductase is specific for the reduction of the disulfide bond of thioredoxin; no other disulfides tested were reduced by the enzyme.[3] On the other hand the thioredoxin *system* (i.e., the combination of thioredoxin reductase, thioredoxin, and TPNH) brings about the reduction of many different disulfides (Fig. 2.3). In this type of reaction thioredoxin-$(SH)_2$ reduces nonenzymatically sterically accessible S-S bonds (e.g., in glutathione, lipoic acid, oxytocin, or insulin). The occurrence of enzyme systems capable of reducing S-S bonds in peptides and proteins was described earlier.[9–12] It is possible that the thioredoxin system is similar to or identical with, for instance, the protein disulfide reductase of Nickerson and Falcone.[9]

Thioredoxin. The key component of the hydrogen transport system is thioredoxin, a relatively heat-stable, acidic protein. We have worked out a simple and reproducible purification procedure that, after about a 6000-fold purification, gives 2–3 μmoles of pure thioredoxin from 1 kg of *E. coli*.[2,13] Some of the properties of thioredoxin are summarized in Table 2.1. The protein contains all common amino acids and has a molecular weight of about 12,000. Its two half-cystine residues form an intrachain S-S bridge, which represents the active center of the molecule.

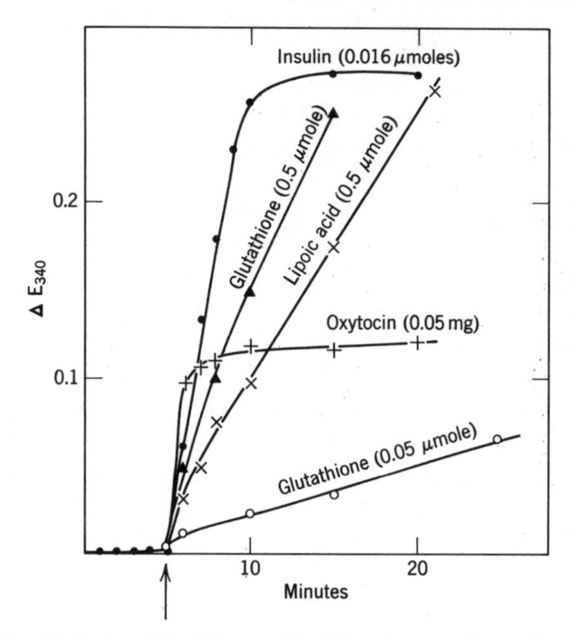

Fig. 2.3. Reduction of different disulfides by complete thioredoxin system.[3] Reaction was started at the arrow by addition of thioredoxin and was followed by measuring the disappearance of TPNH.

Only two amino acids (glycine and proline) are located between the half-cystines (see below). The S-S bridge is presumably located at the surface of the molecule.

No metals or other nonprotein constituents have been found in thioredoxin. This is in contrast to other "re-doxins" that contain nonheme iron, inorganic sulfide, or a flavin (ferredoxin,[14] rubredoxin,[15] or flavodoxin[16]).

The general structure of thioredoxin is given in Fig. 2.4. The molecule consists of a single polypeptide chain of 108 amino acids. It contains a single methionine residue located at position 37, which makes it possible to split the

TABLE 2.1.

Properties of Thioredoxin

Molecular weight	
from ultracentrifugation	12,400
from amino acid composition	11,600
from enzyme activity	11,800
Isoelectric point around 4.4	
The molecule consists of a single polypeptide chain of 108 amino acids with NH_2-terminal serine and COOH-terminal alanine and contains only one residue each of histidine, arginine, cystine and methionine	
All sulfur in the molecule is accounted for by the presence of cystine and methionine	

molecule at this point with cyanogen bromide.[13] Chromatography on Sephadex G-50 with 50% acetic acid results in the separation of the N-terminal peptide B (37 amino acids) and the C-terminal peptide A (71 amino acids).

After further fragmentation of each peptide with different proteolytic enzymes, Holmgren determined the complete amino acid sequence of thioredoxin from the amino acid sequences of the resulting overlapping smaller peptides.[17–19] The sequences for peptides A and B are given in Figs. 2.5 and 2.6, respectively. The amino acid sequence of thioredoxin is obtained by adding the sequence of peptide A to that of peptide B. Methionine should then, of course, be substituted for the C-terminal homoserine of peptide B.

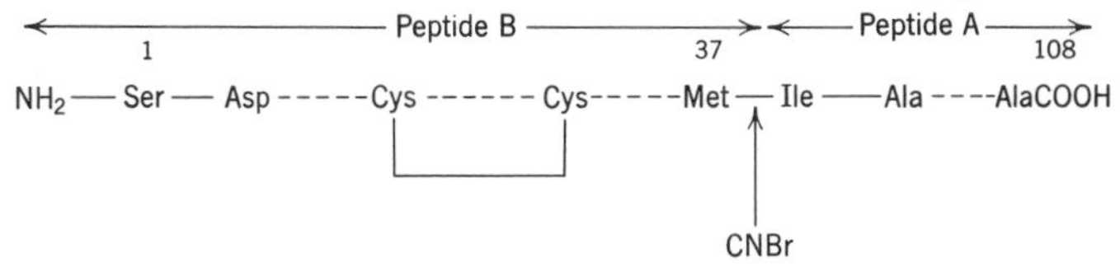

Fig. 2.4. General structure of thioredoxin.

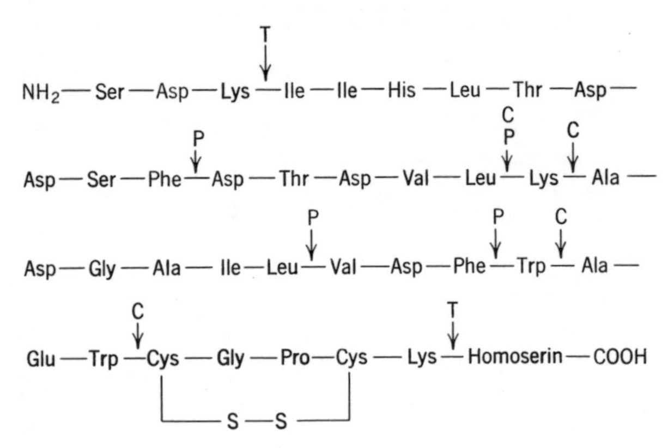

Fig. 2.5. Amino acid sequence of peptide B.[18] Arrows indicate the points of cleavage by trypsin (T), pepsin, (P), and chymotrypsin (C).

Of particular interest is the amino acid sequence around the active center (residues 31–37). A model of this part of the molecule (Fig. 2.7) clearly shows the exposed disulfide bridge that is held in a fixed position by the proline residue

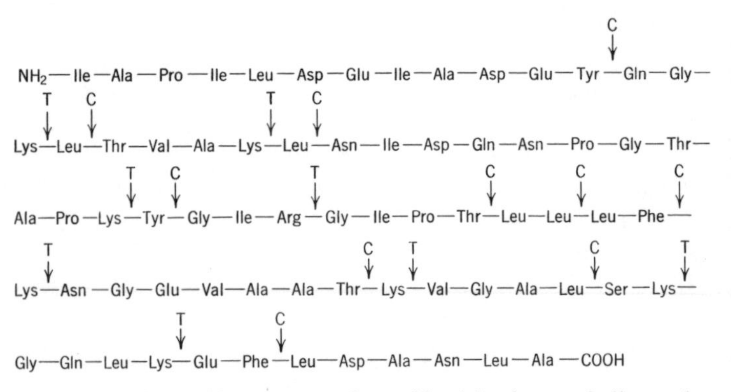

Fig. 2.6. Amino acid sequence of peptide A.[19] Arrows indicate the points of cleavage by chymotrypsin (C) and trypsin (T).

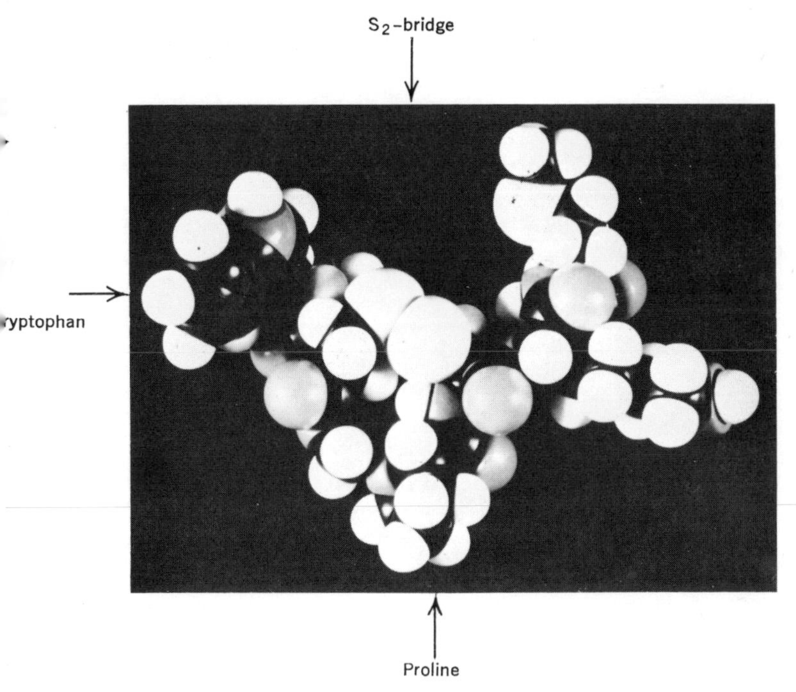

Fig. 2.7. Model of active center of thioredoxin.

next to one of the half-cystines. The close proximity of one of the tryptophans to the disulfide bridge is also evident.

This part of the structure of thioredoxin corroborates in a gratifying way earlier conclusions drawn from experiments that involved the fluorescence of oxidized and reduced thioredoxin.[20] On excitation at 280 mμ (the absorption maximum of tryptophan), oxidized and reduced thioredoxin give the emission spectra shown in Fig. 2.8. It is evident that on reduction of the disulfide bond, the quantum yield of the emission increased almost threefold—

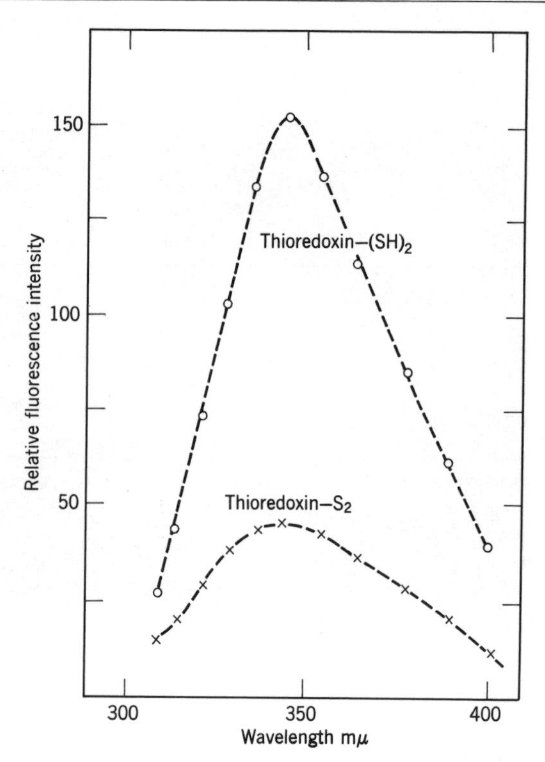

Fig. 2.8. Emission spectra of oxidized and reduced thioredoxin.[20]

indicating that a change had occurred in the conformation of thioredoxin that involved one or both tryptophan residues. In both cases the emission maxima were at 342 mμ, indicating that the tryptophans were located in a highly polar environment.

These results could be explained in two alternative ways:

1. The conformational change signaled by the increased fluorescence of tryptophan involves the whole molecule of

thioredoxin. In this case the tryptophan residues might be located anywhere within the polypeptide chain.

2. During reduction of thioredoxin the conformational change is *localized* to a limited part of the molecule involving the tryptophan residue(s). This would necessitate a close proximity of the tryptophan(s) to the disulfide bridge.

We attempted to distinguish between these two possibilities by comparing several physicochemical parameters (sedimentation behavior, fluorescence polarization, and optical rotatory dispersion) of oxidized and reduced thioredoxin (Table 2.2). We would expect one or several of

TABLE 2.2.

Comparison of Physicochemical Parameters of Oxidized and Reduced Thioredoxin

	Thioredoxin-S_2	Thioredoxin-$(SH)_2$
Sedimentation rate ($s_{20,w}$)	1.36×10^{-13}	1.39×10^{-13}
Fluorescens polarization (p)	0.256 ± 0.003	0.253 ± 0.003
Optical rotatory dispersion $[M']$	-2050 ± 100	-2250 ± 100

these parameters to change on reduction of thioredoxin if an *overall* conformational transition had occurred. As this was not the case, we concluded instead that the opening of the disulfide bridge of thioredoxin results in a *localized* conformational change around the disulfide bridge and the tryptophan residues.

The primary structure of peptide B (Figs. 2.5 and 2.7) confirms this concept very clearly: it shows that both tryptophan residues are indeed localized very close to the disulfide bridge.

Is the localized conformational change around the disulfide bridge during the reduction of thioredoxin connected with the function of thioredoxin as a hydrogen carrier during ribonucleotide reduction? We believe this to be the case. During the hydrogen transport process the reduced form of thioredoxin should be bound strongly to ribonucleotide reductase while the oxidized form should be preferentially bound to thioredoxin reductase. Binding of thioredoxin must in both cases involve the area around the disulfide (or dithiol) group. It seems reasonable to assume that the conformational difference between oxidized and reduced thioredoxin facilitates preferential binding of the two forms to either ribonucleoside diphosphate reductase or thioredoxin reductase.

In extracts from *E. coli* both thioredoxin and thioredoxin reductase apparently occur in excess over the rest of the ribonucleoside diphosphate reductase system. This raises the question whether the thioredoxin system participates as electron carrier system in other reductive processes. I have already mentioned the possibility of its function as a protein disulfide reductase. Other enzyme reactions in which reduced thioredoxin or a compound similar to it may function as reducing agent include the formation of dideoxyhexoses,[21] the reduction of sulfate to sulfite,[22] and the reduction of methionine sulfoxide to methionine.[23] Further possibilities are the formation of cobamide coenzyme from vitamin B_{12}[24] and the reductive deamination of glycine.[25] Finally, a thioredoxin system may also participate in the oxidative decarboxylation of glycine.[26]

Such enzyme reactions appear to have one or several characteristics in common with the reduction of ribonucleotides: the hydrogen transport system involves a pyridine nucleotide and a flavoprotein; and a (di)thiol and reduced lipoic acid or dithioerythrol usually can function as artificial hydrogen donors.

When one looks for "thioredoxin systems" it is important to bear in mind that in most instances no cross reactions occur between thioredoxins and thioredoxin reductases from different organisms. Thioredoxin systems have been demonstrated in *E. coli, L. leichmannii*,[27] Novikoff hepatoma,[28] and yeast.[29] In all instances except one, the reduction of a given thioredoxin was catalyzed only by the thioredoxin reductase obtained from the same organism. The only cross reaction observed so far occurred between thioredoxin reductase from *L. leichmannii* and thioredoxin from *E. coli*.[27]

On the other hand, apparently any reduced thioredoxin —irrespective of the source—can be used as hydrogen donor by different ribonucleotide reductases. A cross reaction was first observed between reduced thioredoxin from *E. coli* and ribonucleotide reductase from Novikoff hepatoma,[30] and later also between *E. coli* thioredoxin and the ribonucleotide reductase from *L. leichmannii*.[31,32] Furthermore, *E. coli* ribonucleotide reductase can use hepatoma[28] or yeast[29] thioredoxin as hydrogen donor. If these findings can be generalized, they may be of considerable practical importance. The ribonucleotide reductase system from, for instance, *E. coli* could then be used as an assay system for the purification of thioredoxin systems from other organisms.

Mechanism of Hydrogen Transfer

Enzyme from E. coli. The formation of deoxyribonucleotides from ribonucleotides involves the transformation of a hydroxymethyl group to a methylene group. From a mechanistic point of view we can ask the question: How is hydrogen from the dithiol of thioredoxin-$(SH)_2$ used for the reduction of the OH-group at position 2′ of the ribonucleotide?

When the enzyme reaction is carried out in ^3H_2O, tritium instantaneously equilibrates with the hydrogens of the dithiol of reduced thioredoxin. During the reduction of the ribonucleotide, tritium is then introduced into deoxyribose from the dithiol. The localization of the isotope within the sugar gives certain indications about the mechanism of the reduction. Thus ^3H would be expected to show up at positions 2′ and 3′ if an intermediate containing a double bond between carbon atoms 2′ and 3′ were involved (second mechanism of Fig. 2.9). On the other hand, if ^3H is introduced exclusively into position 2′, it

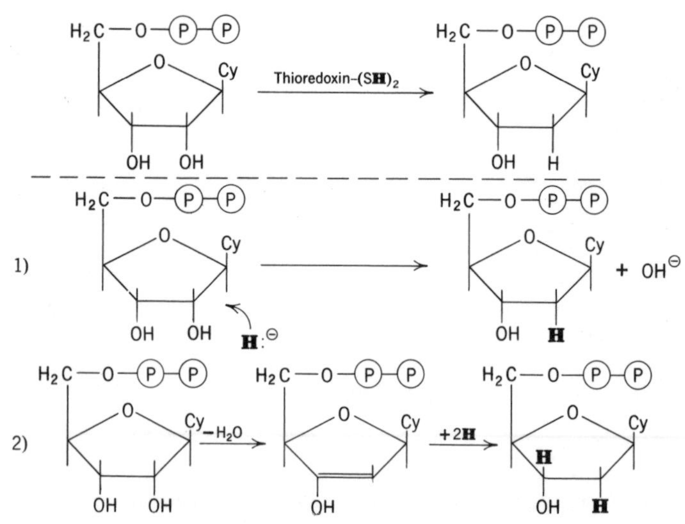

Fig. 2.9. Hypothetical steps in ribonucleotide reduction.[33]

would indicate a direct replacement of the OH-group by hydrogen.

The localization of tritium in dCDP formed by the *E. coli* ribonucleoside diphosphate reductase was determined by Larsson,[33] who worked out suitable methods for the degradation of the deoxyribose from dCDP. The *E. coli* ribonucleoside diphosphate reductase incorporated tritium into position 2′ of the deoxyribonucleotide indicating that the first mechanism of Fig. 2.9 was the correct one.

The reaction showed a considerable isotope effect and only about 0.3 mole of tritium was introduced per mole of dCDP formed. This made it impossible to decide with certainty whether one or two hydrogens were incorporated into deoxyribose. This difficulty was overcome in a different experiment in which 99.7% D_2O was used during the enzyme incubation and the deuterium incorporated into dCDP was localized by NMR spectroscopy.[34] The result not only gave a clear-cut demonstration that only one atom of deuterium was introduced at position 2′—and at no other position—but also gave some clues about the stereochemical course of the reaction. It was evident that the reaction was stereospecific. The incorporation appeared to occur with retention of the stereochemistry at position 2′, that is, deuterium was introduced trans to cytosine. This conclusion rested on earlier calculations by Lemieux[35] concerning the NMR spectra of deoxyribonucleosides.

Enzyme from L. leichmannii. Before considering the reaction mechanism of the cobamide-dependent ribonucleotide reductase from *Lactobacilli,* I shall discuss briefly the mechanism of the reaction catalyzed by the related enzyme dioldehydrase.[36] This enzyme catalyzes a cobamide coenzyme-dependent transformation of diols to aldehydes, for example, the conversion of 1,2 propandiol to propionaldehydé (Fig. 2.10). This reaction is related to ribonucleotide reduction in that it involves the reduction of a CHOH group. In contrast to ribonucleotide reduction, no external reductant—such as reduced thioredoxin—is required because the reaction is an intramolecular oxidoreduction. Of particular interest is the requirement for cobamide coenzyme.

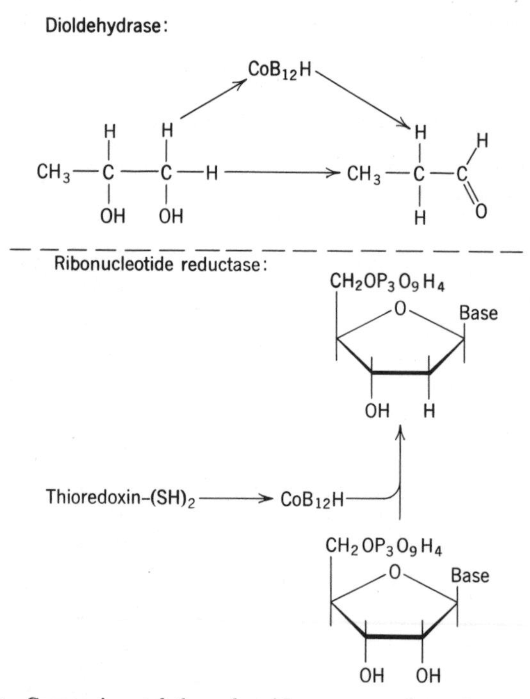

Fig. 2.10. Comparison of the cobamide coenzyme-dependent reductions of diols and ribonucleotides.

The mechanism of this reaction was clarified by the elegant work of Abeles and coworkers.[37–39] They demonstrated that the oxidoreduction is not strictly intramolecular and that cobamide coenzyme participates as a hydrogen carrier that accepts the hydrogen from C-1 of one molecule of propandiol and transfers it to C-2 of a second molecule.

The 5′-methylene of the adenosine moiety of the coenzyme participates in the hydrogen transport. When cobamide coenzyme was labeled with ^{3}H specifically at this position, all isotope was recovered in the propionaldehyde formed.

Let us now return to the ribonucleotide reductase from *L. leichmannii.* Again, cobamide coenzyme functions as a hydrogen carrier—but hydrogen is now accepted by the enzyme-bound coenzyme from reduced thioredoxin and then used for the reduction of the $\overset{|}{\underset{|}{\mathrm{C}}}\mathrm{HOH}$ group of the ribonucleotide (Fig. 2.10).

When ^{3}H-labeled cobamide coenzyme was used with the ribonucleotide reductase, no isotope was incorporated into the deoxyribonucleotide. Instead, the enzyme catalyzed a rapid exchange of isotope between 5′-cobamide-^{3}H coenzyme and water.[40,41] This finding can be explained by the scheme given in Fig. 2.11, if it is assumed that the back reaction between thioredoxin-S_2 and reduced coenzyme is very rapid compared with the reduction of the ribonucleotide. In this way the isotope from the labeled coenzyme would equilibrate with water via the dithiol of reduced thioredoxin.

The mechanism of hydrogen transfer and the stereochemical course of the reaction catalyzed by the cobamide-dependent reductase was studied by Blakley and by Beck and their coworkers[42–44] by methods similar to those men-

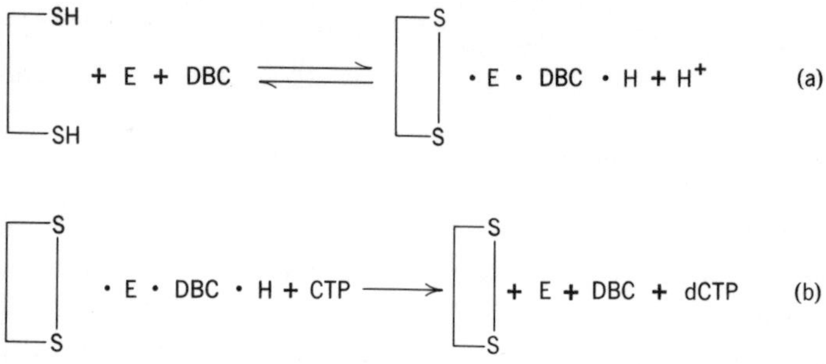

$$\begin{bmatrix} SH \\ \\ SH \end{bmatrix} + E + DBC \rightleftharpoons \begin{bmatrix} S \\ | \\ S \end{bmatrix} \cdot E \cdot DBC \cdot H + H^{+} \qquad (a)$$

$$\begin{bmatrix} S \\ | \\ S \end{bmatrix} \cdot E \cdot DBC \cdot H + CTP \longrightarrow \begin{bmatrix} S \\ | \\ S \end{bmatrix} + E + DBC + dCTP \qquad (b)$$

Fig. 2.11. Function of cobamide coenzyme in ribonucleotide reduction.[40]

tioned for the study of the *E. coli* system. Their results clearly indicate the complete similarity in this respect of the reactions catalyzed by the two ribonucleotide reductases —despite the fact that cobamide coenzyme participates only in the *L. leichmannii* system.

Figure 2.12 compares the intermediate steps involved in the two hydrogen transport systems. In both cases hydrogen is first transferred from TPNH via a flavoprotein to a dithiol (reduced thioredoxin). In both cases the mechanism of the reduction of the ribosyl moiety involves a replacement of the OH-group at position 2′ by a hydrogen, without change in the stereochemistry at this carbon atom; it therefore seems reasonable to propose that the reaction proceeds via a hydride mechanism with the intermediate formation of an enzyme-bound carbonium ion. The major difference between the two enzyme systems concerns the participation of cobamide coenzyme. In *L. leichmannii* the coenzyme appears to function as an intermediate between dihydrothioredoxin and the ribonucleotide; thus according to the scheme of Fig. 2.12 it could generate the required "hydride ion." In the system from *E. coli* this coenzyme does not participate. On the other hand, we recently found that iron forms an integral part of this ribonucleotide reductase.[45] I should like to suggest that this protein-bound nonheme iron generates the "hydride ion" required in the reaction catalyzed by the *E. coli* reductase and that its function thus corresponds to that of cobamide coenzyme in the *L. leichmannii* system. A second but not necessarily alternative hypothesis for the function of nonheme iron might be that it participates in the removal of the hydroxyl group from position 2′ of the ribose.

Mammalian Ribonucleotide Reductases. The enzymes from Novikoff hepatoma and from regenerating rat liver have not yet been obtained in a sufficiently pure state to allow a study of the reaction mechanism along lines similar to those used for the two microbial systems. Neither mammalian system requires the addition of a cobamide deriva-

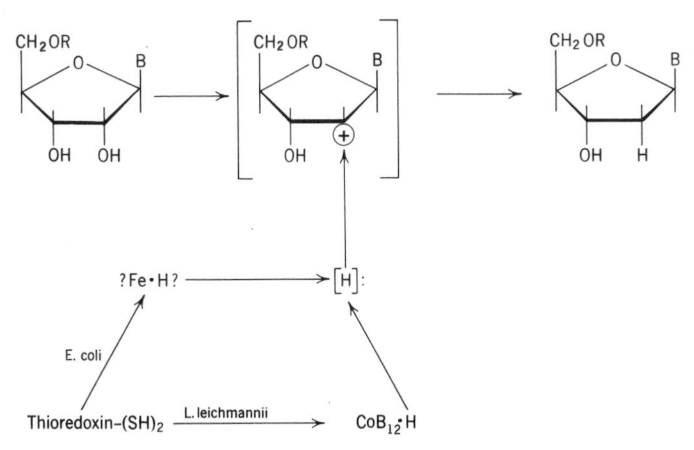

Fig. 2.12. Comparison of intermediary steps during ribonucleotide reduction in *E. coli* and *L. leichmannii.*

tive for enzyme activity. It seems significant that both enzyme preparations are stimulated by Fe^{2+} or Fe^{3+}. The purest preparations of the tumor enzyme actually are almost completely dependent on the addition of iron.[46] In view of the occurrence of iron in the *E. coli* reductase, such results suggest that the mammalian enzymes also may be nonheme iron proteins. However, in their case the iron-protein bond appears to be weaker than in the *E. coli* reductase, which usually is not stimulated by the addition of iron.

The effects of hydroxyurea[47] mentioned in the first chapter may also bear on this question. The reductases from *E. coli* and from mammalian sources—but not the *L. leichmannii* enzyme—are inhibited by hydroxyurea. This compound specifically inhibits protein B2 of the *E. coli* ribonucleotide reductase and there are some indications that this effect is connected with the iron content of the protein. These results thus further stress the similarity between the *E. coli* reductase and the mammalian reductases.

REFERENCES

1. Reichard, P., *J. Biol. Chem.,* **236,** 2511 (1961).
2. Laurent, T. C., E. C. Moore, and P. Reichard, *J. Biol. Chem.,* **239,** 3436 (1964).
3. Moore, E. C., P. Reichard, and L. Thelander, *J. Biol. Chem.,* **239,** 3445 (1964).
4. Thelander, L., *J. Biol. Chem.,* **242,** 852 (1967).
5. Thelander, L., *European J. Biochem.,* **4,** 407 (1968).
6. Massey, V., and C. H. Williams, Jr., *J. Biol. Chem.,* **240,** 4470 (1965).
7. Massey, V., T. Hofmann, and G. Palmer, *J. Biol. Chem.,* **237,** 3820 (1962).
8. Massey, V., Q. H. Gibson, and C. Veeger, *Biochem. J.,* **77,** 341 (1960).
9. Nickerson, W. J., and G. Falcone, *Science,* **124,** 318 (1956).
10. Hatch, M. D., and J. F. Turner, *Biochem. J.,* **76,** 556 (1960).
11. Tomizawa, H. H., and Y. D. Halsey, *J. Biol. Chem.,* **234,** 307 (1959).

12. Katzen, H. M., F. Tietze, and D. Stetten, Jr., *J. Biol. Chem.,* **238,** 1006 (1963).
13. Holmgren, A., and P. Reichard, *European J. Biochem.,* **2,** 187 (1967).
14. Mortenson, L. E., R. C. Valentine, and J. E. Carnahan, *J. Biol. Chem.,* **238,** 794 (1963).
15. Lovenberg, W., and B. E. Sobel, *Proc. Natl. Acad. Sci. U.S.,* **54,** 193 (1965).
16. Knight, E., Jr., and R. W. F. Hardy, *J. Biol. Chem.,* **241,** 2752 (1966).
17. Holmgren, A., R. Perham, and A. Baldesten, *European J. Biochem.,* in press.
18. Holmgren, A., *European J. Biochem.,* in press.
19. Holmgren, A., unpublished results.
20. Stryer, L., A. Holmgren, and P. Reichard, *Biochemistry,* **6,** 1016 (1967).
21. Matsuhashi, S., and J. L. Strominger, *J. Biol. Chem.,* **242,** 3494 (1967).
22. Wilson, L. G., T. Asahi, and R. S. Bandurski, *J. Biol. Chem.,* **236,** 1822 (1961).
23. Black, S., E. M. Harte, B. Hudson, and L. Wartofsky, *J. Biol. Chem.,* **235,** 2910 (1960).
24. Vitols, E., G. Walker, and F. M. Huennekens, *Biochem. Biophys. Res. Commun.,* **15,** 372 (1964).
25. Stadtman, T. C., *Arch. Biochem. Biophys.,* **99,** 36 (1962).
26. Baginsky, M. L., and F. M. Huennekens, *Biochem. Biophys. Res. Commun.,* **23,** 600 (1966).
27. Orr, M. D., and E. Vitols, *Biochem. Biophys. Res. Commun.,* **25,** 109 (1966).
28. Moore, E. C., *Biochem. Biophys. Res. Commun.,* **29,** 264 (1967).
29. Gonzales, P., A. Baldesten, and P. Reichard, unpublished results.
30. Moore, E. C., and P. Reichard, *J. Biol. Chem.,* **239,** 3453 (1964).
31. Vitols, E., and R. L. Blakley, *Biochem. Biophys. Res. Commun.,* **21,** 466 (1965).
32. Beck, W. S., M. Goulian, A. Larsson, and P. Reichard, *J. Biol. Chem.,* **241,** 2177 (1966).
33. Larsson, A., *Biochemistry,* **4,** 1984 (1965).
34. Durham, L. J., A. Larsson, and P. Reichard, *European J. Biochem.,* **1,** 92 (1967).
35. Lemieux, R. U. *Can. J. Chem.,* **39,** 116 (1961).
36. Brownstein, A. M., and R. H. Abeles, *J. Biol. Chem.,* **236,** 1199 (1961).

37. Wagner, O. W., H. A. Lee, Jr., P. A. Frey, and R. H. Abeles, *J. Biol. Chem.*, **241**, 1751 (1966).
38. Frey, P. A., and R. H. Abeles, *J. Biol. Chem.*, **241**, 2732 (1966).
39. Frey, P. A., M. K. Essenberg, and R. H. Abeles, *J. Biol. Chem.*, **242**, 5369 (1967).
40. Abeles, R. H., and W. S. Beck, *J. Biol. Chem.*, **242**, 3589 (1967).
41. Hogenkamp, H. P. C., R. K. Ghambeer, C. Brownson, and R. L. Blakley, *Biochem. J.*, **103**, 5 C (1967).
42. Blakley, R. L., R. K. Ghambeer, T. J. Batterham, and C. Brownson, *Biochem. Biophys. Res. Commun.*, **24**, 418 (1966).
43. Batterham, T. J., R. K. Ghambeer, R. L. Blakley, and C. Brownson, *Biochemistry*, **6**, 1203 (1967).
44. Gottesman, M. M., and W. S. Beck, *Biochem. Biophys. Res. Commun.*, **24**, 353 (1966).
45. Brown, N. C., R. Eliasson, P. Reichard, and L. Thelander, *Biochem. Biophys. Res. Commun.*, **30**, 522 (1968).
46. Moore, E. C., personal communication.
47. Krakoff, I., N. C. Brown, and P. Reichard, *Cancer Res.*, in press.

CHAPTER

3

REGULATORY ASPECTS

Connection with DNA Synthesis. Our long-standing interest in the biosynthesis of deoxyribose is connected with the hope of being able to come to a better understanding of some of the mechanisms involved in the control of DNA synthesis and possibly cell division. The conversion of a ribonucleotide to the corresponding deoxyribonucleotide represents the first step in a reaction sequence that ultimately results in the formation of DNA. All preceding enzyme reactions leading to the synthesis of ribonucleotides have at least a threefold final purpose: the synthesis of RNA, the synthesis of coenzymes, and the synthesis of DNA (Fig. 3.1).

It is well known that branch-points in metabolic sequences are targets of regulatory mechanisms and it could therefore be expected a priori that the reduction of ribonucleotides would be subject to a control mechanism.

The first signs of such a regulatory mechanism were found in 1960 during studies on the synthesis of DNA in

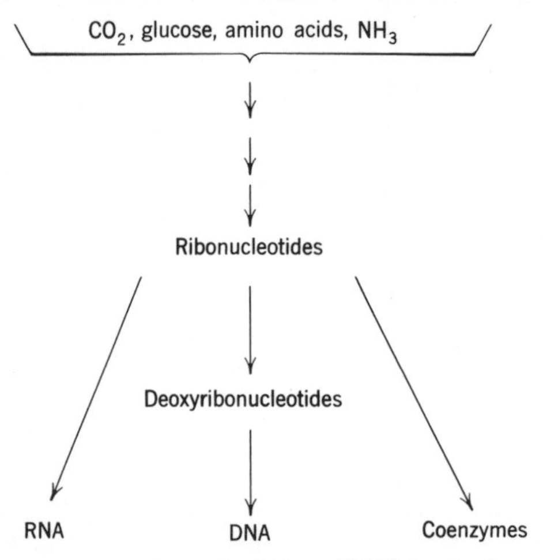

Fig. 3.1. General scheme of DNA synthesis.

extracts from chick embryo.[1] Such extracts carry out a rapid synthesis of DNA from the four deoxyribonucleoside triphosphates (cf. Fig. 1.1). The reaction can be measured by the DNA-dependent formation of an acid insoluble product from, for example, radioactive dCMP in the presence of ATP and Mg^{2+} (Fig. 3.2).

When radioactive CMP is used as substrate for DNA synthesis in place of dCMP, the ribonucleotide must first be reduced to the deoxyribonucleotide (Fig. 3.2). Chick embryo extracts carry out this reduction and thus provide a soluble system for the study of the over-all reaction sequence leading from ribonucleotides to DNA.

The polymerase reaction also requires the three other deoxyribonucleoside triphosphates besides dCTP and it was therefore not surprising to find that an equimolar mixture of dATP, dGTP, and dTTP stimulated the synthesis of DNA from dCMP (Fig. 3.3). On the other hand, the same mixture of deoxyribonucleoside triphosphates showed a pronounced inhibition when radioactive CMP was used as precursor for DNA. This indicated an inhibition of the reductive sequence by one or several of the deoxyribonucleotides.

Such an inhibition could be demonstrated when the reduction of CDP was studied directly without coupling to the polymerase reaction. It was then found that each deoxyribonucleoside triphosphate alone inhibited the reduction of CDP and that in particular dATP showed a profound effect at quite low concentrations.[1]

A slightly different situation developed when radioactive GMP was used as substrate for DNA synthesis. In this case only dATP inhibited the reaction while the two pyrimidine deoxyribonucleoside·triphosphates gave a clear-cut stimulation (Fig. 3.4). Again it was found that the effects of the deoxyribonucleotides were localized to the reductive sequence. Both dCTP and in particular dTTP showed a strong stimulation of the reduction of GDP, while dATP and dGTP were powerful inhibitors.

At that time the general phenomenon of feedback inhibition was well established by the work of Umbarger[2] and Pardee.[3] It seemed likely that the *inhibitory* effects ob-

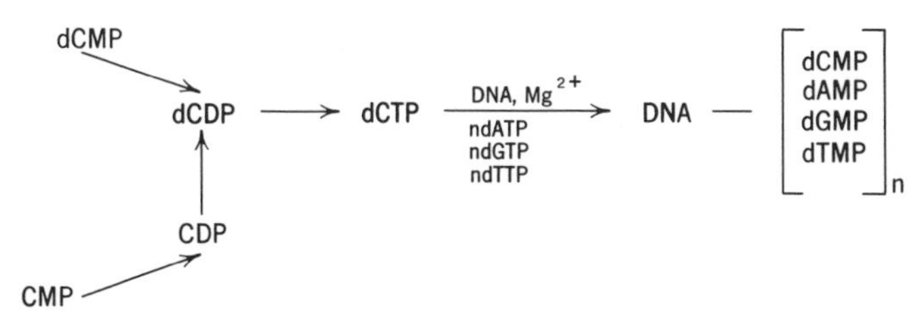

Fig. 3.2. Enzymatic synthesis of DNA from dCMP or CMP.

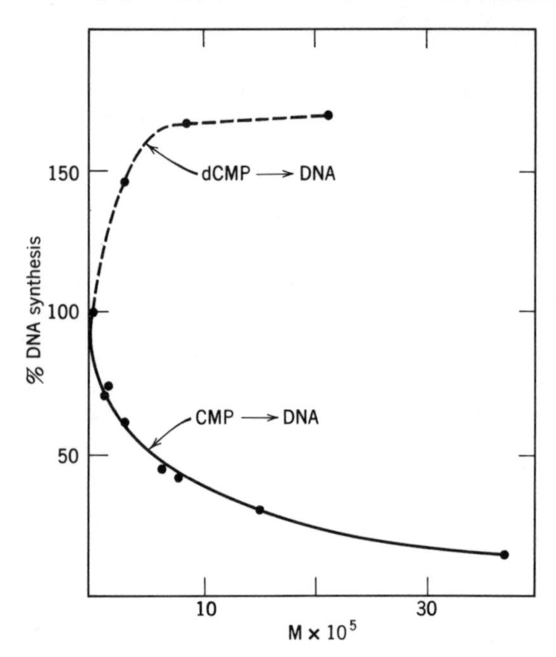

Fig. 3.3. Effect of equimolar mixture of dATP, dGTP, and dTTP on synthesis of DNA from dCMP or CMP.[1]

served in our experiment were of this general nature. The *stimulatory* effects found in the reduction of GDP were, on the other hand, more difficult to understand. We had apparently found a few fragments of a more general mechanism involved in the regulation of the supply of DNA precursors; an understanding of the complete mechanism obviously required studies of less complicated systems. We therefore decided to obtain a highly purified ribonucleotide reductase before attempting to combine the fragments in order to gain a complete picture of the control mechanism.

Chick embryo extract did not appear to be an attractive

source for enzyme purification. Instead, as described in the two previous chapters, we turned our attention to microorganisms and over a period of years purified and characterized an enzyme system from *E. coli* B which in the presence of ATP, Mg^{2+}, and the thioredoxin system catalyzed the reduction of the two pyrimidine ribonucleoside diphosphates CDP[4] and UDP.[5] It was then possible to reopen the question of the control of ribonucleotide reduction by the use of highly purified enzyme preparations without the complication of interfering enzymes such as kinases and phosphatases.

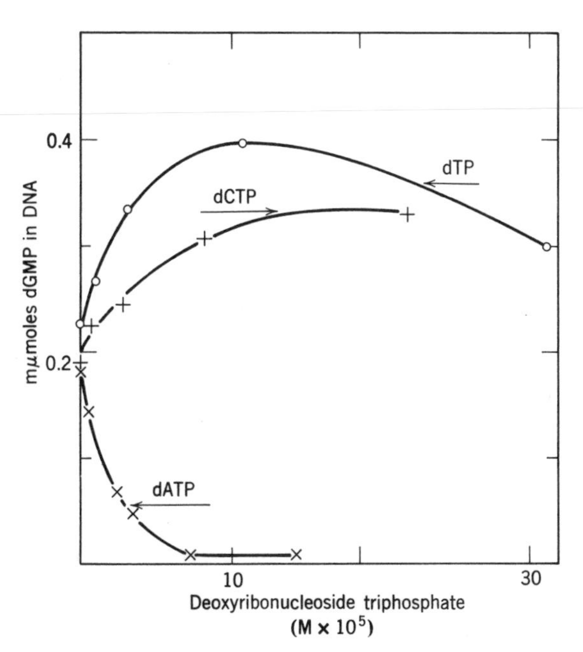

Fig. 3.4. Effect of individual deoxyribonucleoside triphosphates on synthesis of DNA from GMP.[1]

Allosteric Effects. At this stage of our investigations we were puzzled by two questions: (1) What was the substrate specificity of our enzyme?—could it reduce other ribonucleoside diphosphates as well as CDP and UDP, or are there different enzymes for the reduction of pyrimidine and purine ribonucleotides? [4–6] (2) What was the function of ATP in the enzyme system that reduces pyrimidine ribonucleotides?

These two questions somewhat unexpectedly turned out to be connected. Their solution eventually led to an understanding of the control mechanisms involved in ribonucleotide reduction.

Let us start with the ATP requirement. Originally it was thought that ATP, together with Mg^{2+}, was required for an "activation" of the OH-group at position 2′ of the ribonucleotide.[4] If this were the case, one mole of ATP should be consumed for each mole of dCDP formed during the reduction of CDP. However, with the purified enzyme we could easily show that all ATP remained unchanged during the course of the enzyme reaction.[7] This effectively ruled out our original hypothesis.

Furthermore, the requirement for ATP was not absolute and the degree of stimulation by ATP was dependent on the concentration of CDP—being much more pronounced at low concentrations of substrate. A detailed analysis revealed that the main effect of ATP was to decrease the apparent K_m value for CDP by almost one order of magnitude. Furthermore, at saturating concentrations of CDP, the V_{max} of the reaction was increased more than twofold. These results strongly suggested that we were dealing with allosteric effects and that ATP acts as a positive effector in the system.[7]

Other nucleoside triphosphates also showed pronounced effects on the reduction of CDP, acting either as positive or negative allosteric effectors. Thus dTTP stimulated CDP reduction to the same extent as did ATP. Maximal stimulation was obtained at much lower concentrations. On the other hand, dATP strongly inhibited the reaction.[7,8]

The simultaneous addition of two effectors gives rather complex results. A clear-cut antagonism between the positive effects of ATP and the inhibitory effects of dATP is found. At any concentration of dATP, increasing amounts of ATP can overcome the inhibition and, as demonstrated in Fig. 3.5, addition of ATP to dATP inhibited systems

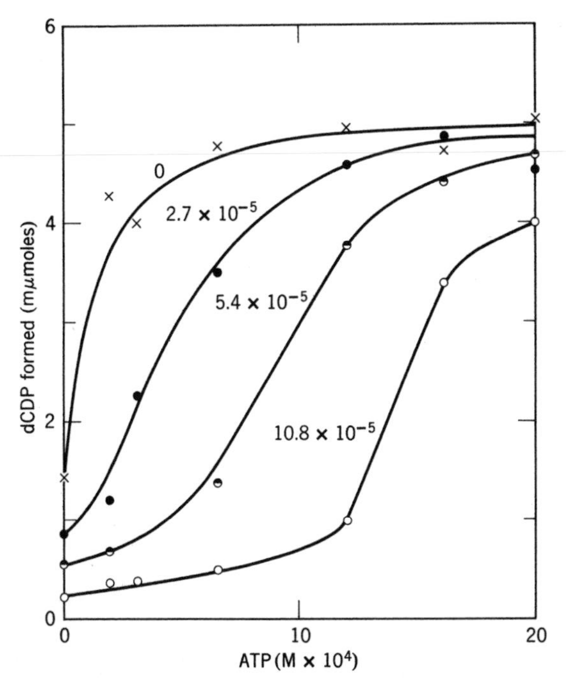

Fig. 3.5. Reversal of dATP-inhibition by ATP.[7] The numbers with each curve give the different concentrations of dATP.

results in a series of S-shaped curves.[7] This is the only instance in which we found the sigmoidal kinetics usually considered typical for allosteric systems. In all other instances normal Michaelis-Menten kinetics are observed.

In contrast to ATP, the positive effector dTTP does not counteract the inhibition by dATP.[8] Actually, in this situation dTTP increases the inhibitory effects of dATP. A still more paradoxical situation occurs when the two positive effectors, ATP and dTTP, are added simultaneously. This usually causes a quite pronounced inhibition of the enzyme reaction. No explanation can be offered for these results.

Nevertheless, it is clear that ATP—and several other nucleoside triphosphates—act as allosteric effectors[9] in the reduction of CDP. Let us now return to the other question posed originally, that of the substrate specificity of the enzyme.

Substrate Specificity. In crude extracts of *E. coli* pyrimidine and purine ribonucleotides are reduced with equal effectiveness.[6] During the course of enzyme purification—using the reduction of CDP as an assay method—the activity toward UDP remained unchanged while that toward purine ribonucleoside diphosphates was lost to a large extent. It therefore seemed reasonable to assume that separate enzymes are involved in the reduction of purine and pyrimidine ribonucleotides.[6] In all these early experiments the substrate specificity of the enzyme was investigated under conditions found optimal for the reduction of CDP, i.e., in the presence of ATP.

When it became clear that ATP acts as an allosteric effector, it seemed possible that different positive effectors are required for the reduction of pyrimidine and purine ribonucleotides and that this might explain our earlier negative results with purine ribonucleotides. The earlier-mentioned stimulatory effect of dTTP on the reduction of GDP by chick embryo extracts[1] was an indication in this direction.

When dTTP was used as effector instead of ATP, the enzyme from *E. coli* catalyzed a rapid reduction of purine ribonucleotides.[10] The earlier negative results with ATP depended on the fact that this nucleotide is not a positive effector. Of the other nucleoside triphosphates, dGTP shows a positive effect similar to that of dTTP, while dATP is a strong negative effector. Figure 3.6 illustrates these effects on the reduction of GDP.

When ADP is used as substrate the results are similar to those obtained with GDP. The only difference is that dTTP

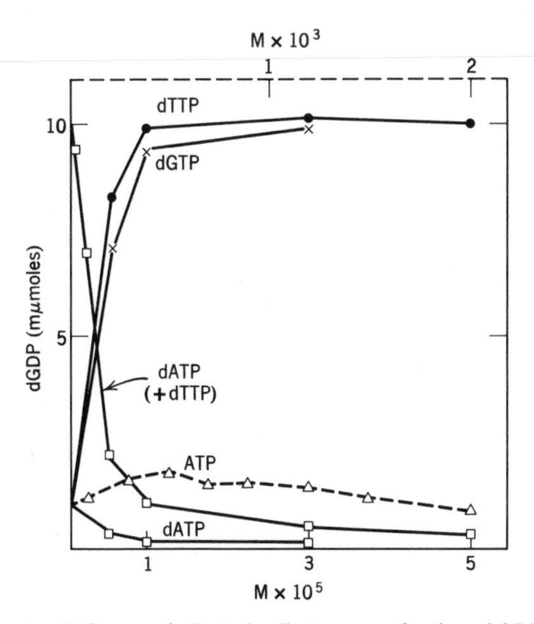

Fig. 3.6. Influence of allosteric effectors on reduction of GDP.[10]

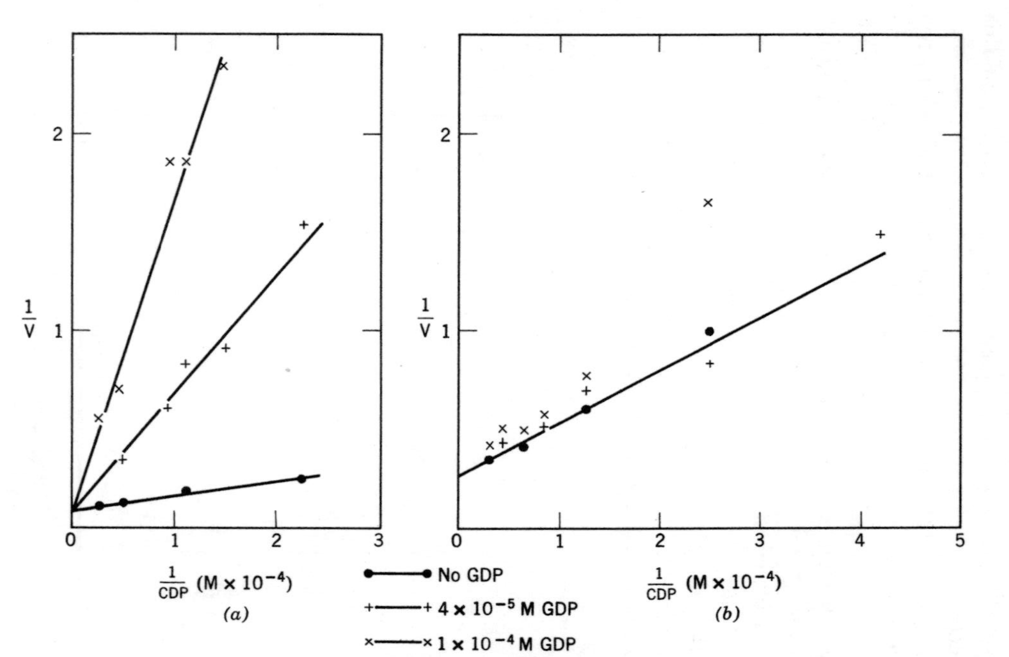

Fig. 3.7. Competition between GDP and CDP for ribonucleotide reductase.[8] dTTP was used as effector in experiment shown in left diagram (*a*); ATP in right diagram (*b*).

is a slightly better positive effector for reduction of GDP, while dGTP gives the best results with ADP. Again dATP is a strong negative effector and the inhibition by this nucleotide is counteracted by ATP—despite the fact that ATP alone does not influence the reduction of either ADP or GDP.

These results clearly demonstrate that a highly purified preparation of ribonucleotide reductase from *E. coli* has the capacity to reduce both purine and pyrimidine ribonucleotides[10]—they do not prove that the same enzyme protein is involved in both types of reaction. However, there is good reason to believe that the same enzyme indeed catalyzes the reduction of all four ribonucleoside diphosphates. First, during enzyme purification the activities toward all four substrates were purified in parallel during the purifications of both protein B1 and protein B2. Second, a clear-cut competition occurs between different substrates provided the proper allosteric effector is used. Thus with dTTP as effector the reduction of CDP is inhibited by increasing amounts of GDP. The inhibition is mutual, i.e., CDP inhibits in the same experiment the reduction of GDP. Lineweaver-Burk plots (Fig. 3.7*a*) indicate a typical competitive situation with the K_i values for the two substances being very close to their corresponding K_m values. On the other hand, when ATP is substituted for dTTP as effector, addition of increasing amounts of GDP influences the reduction of CDP hardly at all (Fig. 3.7*b*). It should be recalled that ATP acts as positive effector only in the reduction of CDP.

We conclude from these experiments that the substrate specificity of the ribonucleoside diphosphate reductase is determined by allosteric effectors. The enzyme system can occur in three alternative active states: (1) in the presence of ATP the enzyme assumes a pyrimidine-specific state; (2) in the presence of dGTP, a purine specific state; and (3) in

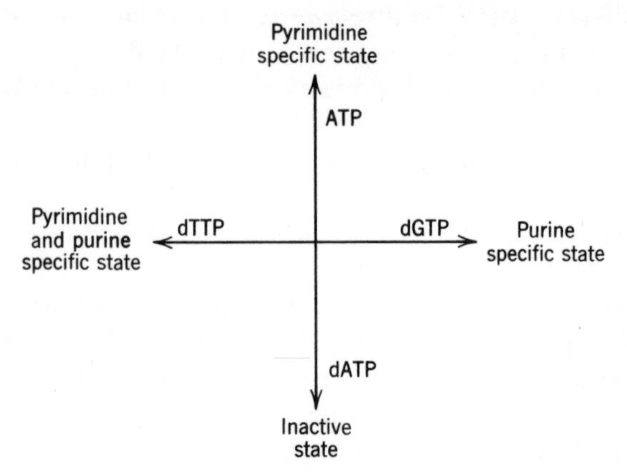

Fig. 3.8. Allosteric effects on ribonucleoside diphosphate reductase from *E. coli*.[10]

the presence of dTTP, a state active with both pyrimidine and purine ribonucleotides. Finally, in the presence of dATP there exists a fourth state in which the enzyme shows little catalytic activity. These interrelationships are summarized in Fig. 3.8.

Subunit Structure. Thus far we have not yet considered the finding that the ribonucleotide reductase from *E. coli* consists of two proteins, proteins B1 and B2. I mentioned earlier that the simultaneous presence of both is required for enzyme activity: each protein alone is completely inactive and does not catalyze any known partial reaction.

I will now present evidence to show that proteins B1 and B2 are nonidentical subunits of the enzyme and that Mg^{2+} is required for the binding of the subunits.[11]

In a buffered 5–20% linear sucrose gradient containing dithiothreitol and 0.01 M Mg^{2+}, protein B1 moves as a single component with a sedimentation coefficient of about 7.8 S, while protein B2 alone shows a value of 5.5 S (Fig. 3.9). When both proteins B1 and B2 are centrifuged simultaneously in the presence of Mg^{2+}, a new protein peak appears that then has a sedimentation coefficient of about 8.8 S.

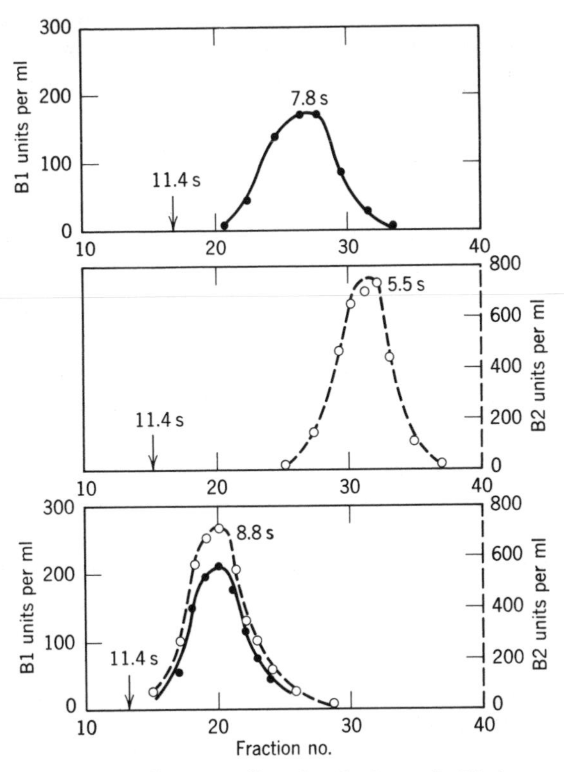

Fig. 3.9. Sucrose gradient centrifugations[11] of protein B1 (upper curve), protein B2 (middle curve), and a mixture of B1 plus B2 (lower curve) with 0.01 M $MgCl_2$. Catalase (11.4 S) was used as a marker.

This peak contains both B1 and B2 and the two activities coincide quite well. The appearance of this new peak is dependent on the presence of Mg^{2+} in the gradient.

The experiments illustrated in Fig. 3.9 were carried out with a slight excess of protein B1 over protein B2. Figure 3.10 illustrates a second experiment in which different proportions of proteins B1 and B2 were centrifuged simultaneously in the presence of Mg^{2+}. In the experiment of the top row, protein B1 is present in excess over protein B2. The middle and bottom rows illustrate two experiments with increasing amounts of protein B2. While the B1:B2 complex peak appears in all three runs, the peak corresponding to free B2 activity is seen only in the last two experiments—in which protein B2 was present in excess. In this experiment the overlap between B1 and B2 activities in the complex peak is not so good as in the experiment of Fig. 3.9. This indicates a dissociation of the complex during centrifugation. Because of this we cannot determine the true sedimentation coefficient of the complex and do not know with certainty the B1:B2 stoichiometry of the active complex. The combination of one subunit of B1 with one subunit of B2 appears to be most likely from our data.

Complex formation is dependent on the presence of Mg^{2+}. How do allosteric effectors influence this phenomenon? We have just started experiments on this question and have obtained results only with two effectors, dTTP and dATP.[12] In spite of the fact that these results are only of preliminary nature, I want to discuss them briefly because they are of considerable importance for our interpretation of the molecular mechanisms involved in the regulation of ribonucleotide reductase.

We find that dTTP does not influence complex formation, at least as studied by the sucrose gradient centrifugation technique. Thus in the absence of Mg^{2+}, addition of

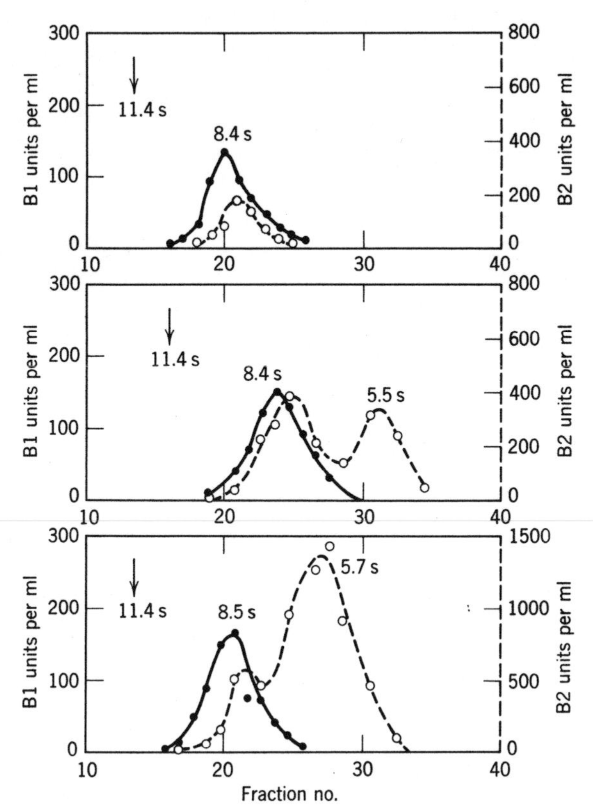

Fig. 3.10. Sucrose gradient centrifugations of different proportions of proteins B1 plus B2 in 0.01 M $MgCl_2$. A constant amount of B1 was centrifuged with increasing amounts of B2. Amount of protein B2 was increased fourfold between each experiment.

dTTP does not result in complex formation, while in the presence of Mg^{2+} a B1:B2 complex is formed with about the same sedimentation behavior as in the absence of dTTP.

More complicated results are obtained with dATP in the

sucrose gradients. It was recently found[13] that very low concentrations (below 10^{-6} M) of this nucleotide stimulate the reduction of CDP. These concentrations of dATP in the gradient give the same results as dTTP: the normal B1:B2 complex is formed in the presence of Mg^{2+}. The picture changes completely when inhibitory concentrations of dATP are used. In the experiment illustrated by Fig. 3.11 dATP (8.8×10^{-5} M) was present during the centrifugation of a mixture of B1 and B2, with the latter protein being in excess. Clearly almost all of the B1 activity is then present in a heavy peak at around 13.5 S. The larger part of the B2 activity is also associated with this peak. In control experiments it was found that this concentration of dATP did not influence the sedimentation behavior of the separated B1 and B2 proteins.

It thus seems likely that inhibitory concentrations of

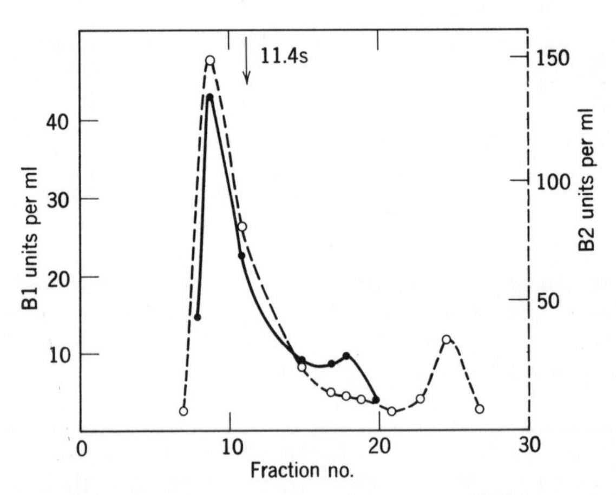

Fig. 3.11. Sucrose gradient centrifugation of proteins B1 and B2 in 0.01 M $MgCl_2$ and 8.8×10^{-5} M dATP. Catalase (11.4 S) was used as marker.

dATP induce the formation of a heavy peak containing both types of subunits. It is tempting to connect the appearance of this "heavy complex" with the inhibitory action of the nucleotide, but direct proof on this point is difficult to obtain.

The discovery of two subunits in ribonucleotide reductase brings to mind other enzymes containing nonidentical subunits, for example, tryptophan synthetase[14] or aspartate carbamyl transferase.[15] The allosteric regulation of the latter enzyme was clarified by the elegant work of Gerhart and Schachman,[15] who demonstrated that one of the two subunits of this enzyme is involved in the actual catalytic process (catalytic subunit) while the other has a regulatory function (regulatory subunit). It appeared possible that a similar division of labor takes places between the two subunits of ribonucleotide reductase.

A regulatory subunit should have the capacity to bind allosteric effectors. We therefore investigated the ability of the separated and combined B1 and B2 proteins to bind dTTP or dATP.[11,12] For this purpose the radioactive nucleotide was first briefly incubated with the protein and the mixture was then filtered through a short column of Sephadex G-25 (Fig. 3.12). The presence of radioactivity in the protein peak (first peak) signals binding of the nucleotide. It is evident that protein B1—either alone or in combination with protein B2—can bind dTTP, while protein B2 alone shows no such ability. Similar results were obtained with dATP.

These results are of a qualitative nature. Quantitative data on the binding of effectors to protein B1 were obtained by the technique of Hummel and Dreyer.[16] In these experiments Sephadex G-50 columns were first equilibrated with different concentrations of the radioactive nucleotide and protein B1 was then filtered through the column. During

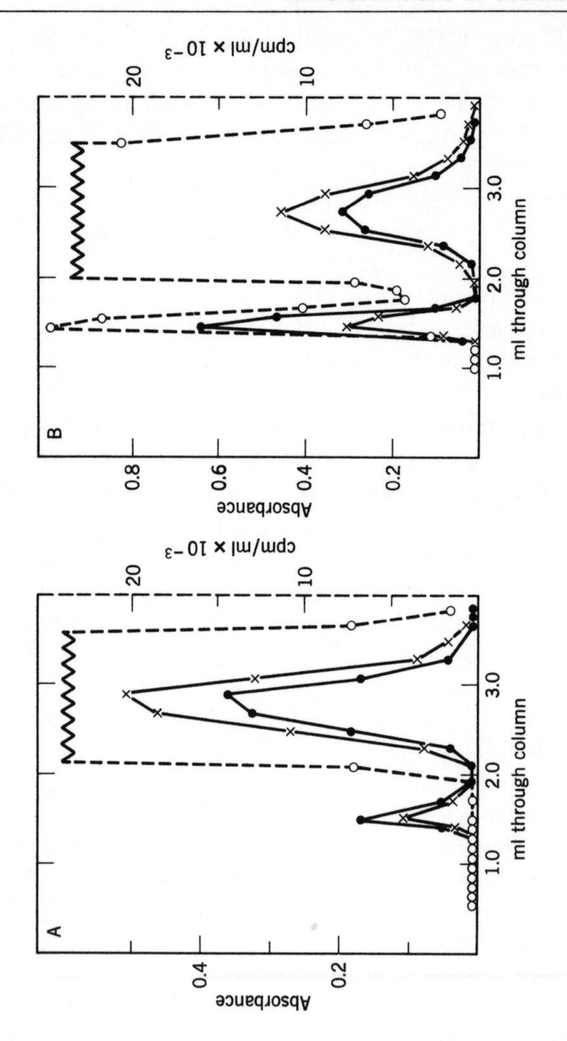
A
Absorbance
0.4
0.2
ml through column
1.0
2.0
3.0
cpm/ml × 10 −3
20
10
B
Absorbance
0.8
0.6
0.4
0.2
ml through column
1.0
2.0
3.0
cpm/ml × 10 −3
20
10

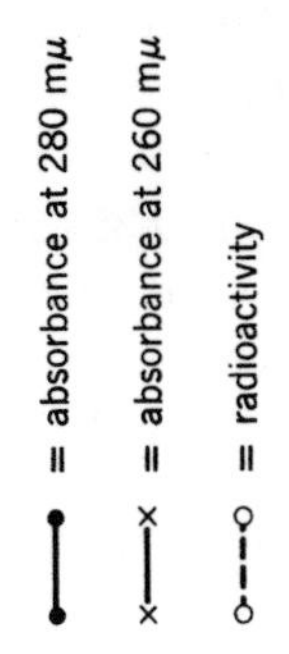

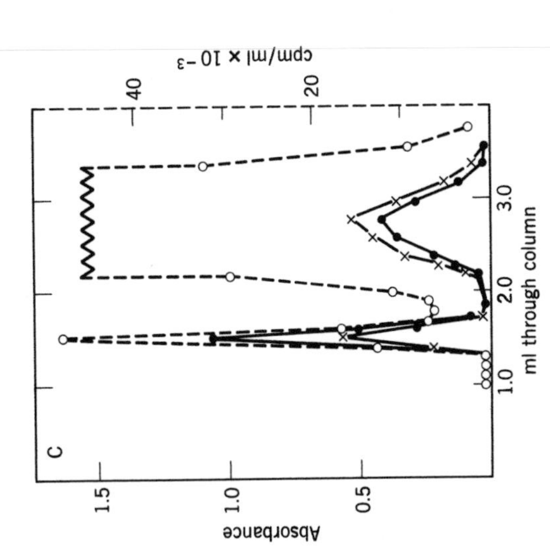

Fig. 3.12. Gel filtration of proteins B1 and B2 on Sephadex G-25 after incubation with ^{3}H-dTTP.[11] Diagram A represents experiment with protein B2; B, with protein B1; C, with B1 plus B2.

chromatography the protein equilibrates with the nucleotide and when the protein peak emerges from the column it is paralleled by a peak of radioactivity (Fig. 3.13). The amount of nucleotide bound per milligram of protein can be determined from the "specific radioactivity" of the protein and is dependent on the equilibrium concentration of the nucleotide in the Sephadex column and the affinity of protein B1 for the nucleotide.

Figure 3.14 gives Lineweaver-Burk plots for the binding of dTTP, dATP, and dGTP. The binding constants for all three nucleotides were slightly over 10^{-6} M and under the conditions of these experiments 1 mg of this preparation of protein B1 bound a maximum of 13–17 mμmoles of nucleotide. As mentioned earlier it was difficult to obtain stable preparations of the protein and these values must be considered to be minimal values. When in a separate

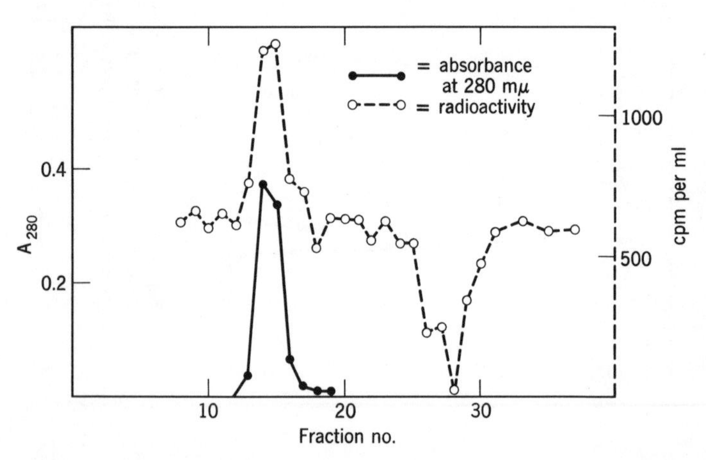

Fig. 3.13. Filtration of protein B1 through column of Sephadex G-50 equilibrated with ^{3}H-dTTP.[16]

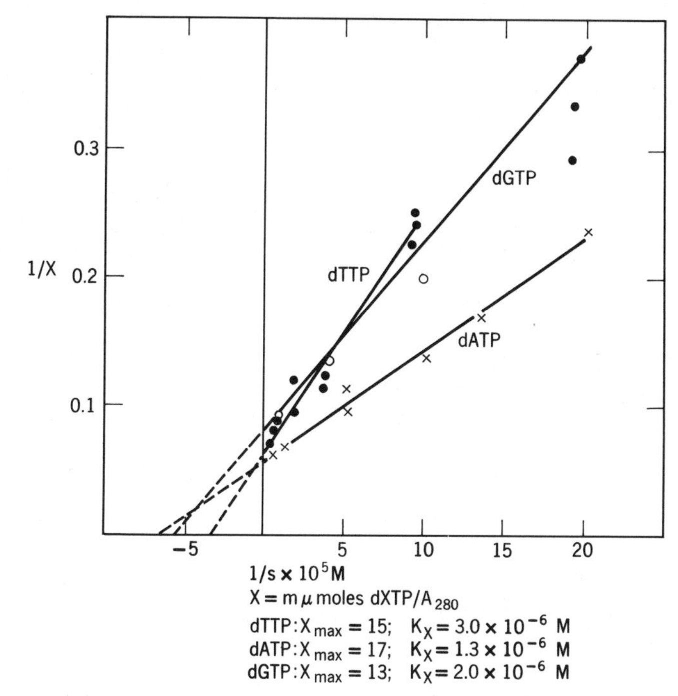

Fig. 3.14. Binding of allosteric effectors to protein B1. All experiments were done with 0.01 M $MgCl_2$ with the technique of Hummel and Dreyer.[16]

experiment special precautions were taken to obtain a stable preparation of protein B1, a total of 19 mμmoles of dATP was bound per milligram of protein. By sedimentation equilibrium centrifugation techniques, the molecular weight of protein B1 was found to be 200,000.[12] With this value one can calculate that 1 mole of this preparation of protein B1 could bind 3.8 moles of dATP, which prob-

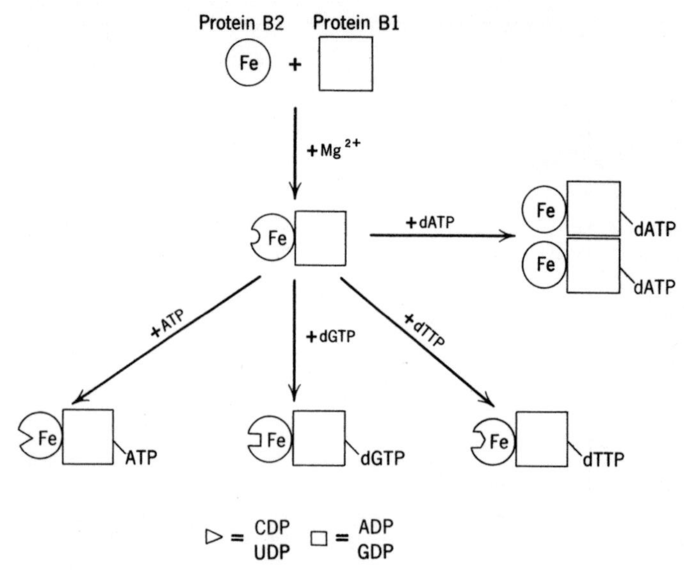

Fig. 3.15. Hypothetical model of different forms of ribonucleoside diphosphate reductase from *E. coli.*

ably indicates that the pure protein can bind 4 molecules of the effector.*

A tentative schematic representation of the enzyme is given in Fig. 3.15. One of the two subunits (protein B1) binds the allosteric effectors.** The binding site for the

* The presence of more than one binding site per subunit of B1 suggests that each subunit contains more than one polypeptide chain. In effect, on centrifugation in 6M guanidinium hychrochloride and 0.1M mercaptoethanol a molecular weight of about 100,000 was obtained, indicating that under these conditions protein B1 dissociated into two polypeptide chains of equal size.

** For the sake of simplicity the scheme does not take into consideration the occurrence of more than one binding site per subunit.

ribonucleotides is given on the other subunit (protein B2), even though there is no direct evidence on this point. However, protein B2 contains iron, which in all probability participates directly in the catalytic process.[17] This is a strong argument for the assumption that protein B2 contains at least part of the catalytic site.

Wherever the exact location of the ribonucleotide binding site may be, it seems clear that the nature of the allosteric effector bound to protein B1 influences the conformation of the active site. Thus in the presence of ATP the site can accommodate a pyrimidine ribonucleotide, with dGTP a purine ribonucleotide and with dTTP either a purine or a pyrimidine substrate. The conformational changes appear to occur without gross changes in the molecular weight of the complex and are schematically indicated in Fig. 3.15.

On the other hand, binding of dATP apparently results in a more complex situation. At very low concentrations (10^{-6} M or less) dATP acts like ATP and stimulates the reduction of pyrimidine ribonucleotides.[13] At concentrations above 10^{-6} M, dATP begins to inhibit the reduction of both pyrimidine and purine ribonucleotides and this general inhibitory effect on ribonucleotide reductase is paralleled by the formation of an aggregated form of the enzyme. While a better understanding of this effect and its connection with the catalytic activity of the enzyme requires further experiments, it seems quite likely that the inhibition of the enzyme by dATP is indeed correlated to the formation of this heavy complex. Our results favor the assumption that this complex contains two subunits each of B1 and B2 (Fig. 3.15).

Thus the allosteric effects appear to operate along two different lines:

1. The regulation of the substrate specificity of the en-

zyme occurs by rather subtle conformational alterations—resulting in a change of the active site of the enzyme. This type of effect appears to be similar in nature to the effect exerted by the regulatory subunit of aspartate carbamyl transferase on the catalytic subunit of the enzyme.[15,18]

2. The inhibition of ribonucleotide reductase by dATP on the other hand appears to be caused by a change in the state of aggregation of the enzyme. This effect on the quaternary structure of the enzyme parallels results described earlier for several other enzymes.[19–23]

Before leaving the subject of the subunit structure of ribonucleotide reductase, I should like to point out one important unsolved question: Which part of the molecule contains the substrate binding sites? Protein B1 appears to fulfill the requirements for a regulatory subunit, and it may therefore be tempting to assign to protein B2 the role of a catalytic subunit, in particular because it contains a cofactor (nonheme iron) required for the catalytic function of the enzyme. However, protein B2 alone has no known catalytic activity. At the present time two alternative explanations appear possible.

1. Protein B2 *is* the catalytic subunit and contains potential binding sites for both thioredoxin and the ribonucleotide. However, these sites only attain the proper configuration for the binding of the substrates when protein B2 is attached to protein B1.

2. Alternatively the catalytic site for the binding of substrates is located in part on protein B2 and in part on protein B1. The complete active site is formed only when the two proteins are combined.

Physiological Implications. The allosteric regulation of highly purified ribonucleotide reductase from *E. coli* shows many of the characteristic features of the regulatory effects found originally with the crude ribonucleotide reductase system from chick embryo and described at the beginning of this chapter. There is also a most striking similarity with the partially purified mammalian system from Novikoff hepatoma studied in detail by Moore and Hurlbert.[24] The scheme given in Fig. 3.16 represents an attempt to integrate the effects observed *in vitro* with *E. coli* reductase into a meaningful physiologic control mechanism for the formation of deoxyribonucleotides.[10] A similar scheme was proposed by Moore and Hurlbert[24] on the basis of their results with the mammalian enzyme.

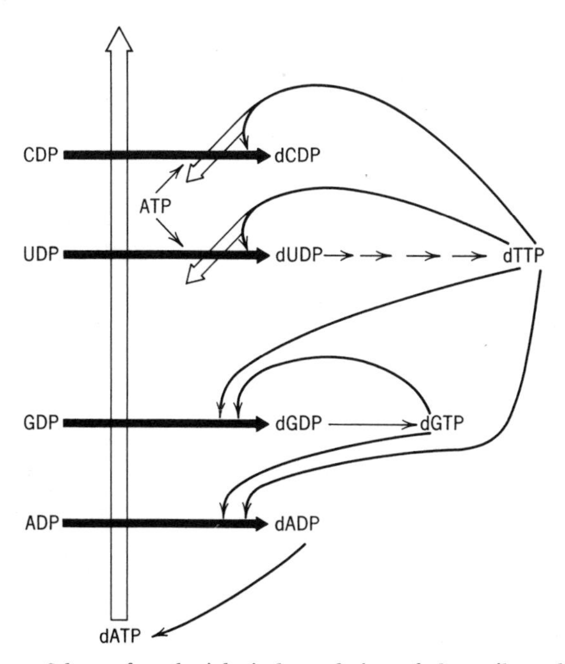

Fig. 3.16. Scheme for physiological regulation of deoxyribonucleotide synthesis in *E. coli*.[10]

The basic features of the scheme are the following: (1) ATP stimulates the formation of pyrimidine deoxyribonucleotides; (2) dTTP and/or dGTP are required for the formation of purine deoxyribonucleotides; and (3) dATP inhibits the formation of all deoxyribonucleotides.

Deoxyribonucleotide formation is visualized as a sequential process: in the first phase, ATP initiates the formation of pyrimidine deoxyribonucleotides. This results, among other things, in the formation of dTTP. This nucleotide then initiates the second phase, the formation of purine deoxyribonucleotides. This process—and in particular the reduction of ADP—is accelerated by the dGTP formed during the reduction of GDP. Finally, in the third phase, dATP starts to accumulate and eventually completely shuts off the enzyme.

The purpose of the whole process is to provide a balanced supply of substrates for DNA polymerase. As long as DNA synthesis takes place, dATP does not accumulate and the reductase will continue to produce deoxyribonucleotides. The allosteric mechanism ensures that in the absence of DNA synthesis no overproduction of deoxyribonucleotides occurs. In line with this concept is the experimental finding that the pools of deoxynucleoside triphosphates—in particular those of the purine compounds—appear to be extremely small in most cells.[25]

Our knowledge of the composition of the acid soluble pools of deoxyribosyl compounds is, however, quite limited. To a large extent this is no doubt caused by the inadequacies of the analytical methods available. Further systematic work in this field comprising, for example, experiments with synchronized bacterial and mammalian cells may be expected to shed some light on a possible connection between the control of DNA synthesis and the regulation of ribonucleotide reductase.

Regulation in Intact Cells. The addition of individual deoxyribonucleosides to whole cell systems often results in the inhibition of cell division, which is caused by an inhibition of DNA synthesis.[26–33] In some cases chromosomal aberrations were also observed.[34]

This type of effect was first demonstrated by Klenow,[26,30] who found that the synthesis of DNA in Ehrlich ascites cells—as measured by the incorporation of labeled inorganic phosphate or formate—could be almost completely stopped by the addition of 2×10^{-3} M deoxyadenosine. Further work[28] demonstrated that deoxyadenosine in these cells was rapidly transformed to dATP and that the inhibition of DNA synthesis by deoxyadenosine could be reversed by the simultaneous addition of deoxyguanosine and deoxycytidine.[30] The most likely interpretation of these results is that the inhibition of DNA synthesis by deoxyadenosine is caused by an inhibition of ribonucleotide reductase by dATP as outlined in Fig. 3.16. The addition of other deoxynucleosides to the medium bypasses the inhibition of the reductase and allows DNA synthesis to proceed.

A similar type of inhibition can be observed with thymidine.[29,31–33] A cell line that is deficient in thymidine kinase and resistant to the inhibition by thymidine could be isolated, indicating that a phosphorylated derivative of thymidine rather than the nucleoside per se was the inhibitor of DNA synthesis.[31] The inhibition was reversed by deoxycytidine in accordance with the scheme of Fig. 3.16, which shows that dTTP only inhibits the reduction of pyrimidine ribonucleotides.

The experiments demonstrating inhibition of DNA synthesis in intact cells by deoxyribonucleosides fit well into the pattern of allosteric effects observed with soluble enzyme systems and may give us confidence that the scheme given in Fig. 3.16 at least represents an approximation to the situation in the living cell. Indirectly, these experiments also allow another important conclusion: inhibition of ribonucleotide reductase clearly leads to a block in the synthesis of DNA; therefore the reduction of ribonucleotides appears to be a prerequisite for DNA synthesis. Short of the existence of mutants blocked in ribonucleotide reduction, these experiments represent the best physiological evidence available that the reduction of ribonucleotides represents an obligatory step in the biosynthesis of DNA.

REFERENCES

1. Reichard, P., Z. N. Canellakis, and E. S. Canellakis, *J. Biol. Chem.*, **236,** 2514 (1961).
2. Umbarger, H. E., *Science,* **123,** 848 (1956).
3. Yates, R. A., and A. B. Pardee, *J. Biol. Chem.,* **221,** 757 (1956).
4. Reichard, P., *J. Biol. Chem.,* **236,** 2511 (1961).
5. Bertani, L. E., A. Häggmark, and P. Reichard, *J. Biol. Chem.,* **238,** 3407 (1963).
6. Larsson, A., *J. Biol. Chem.,* **238,** 3414 (1963).
7. Holmgren, A., P. Reichard, and L. Thelander, *Proc. Natl. Acad. Sci. U.S.,* **54,** 830 (1965).
8. Larsson, A., and P. Reichard, *J. Biol. Chem.,* **241,** 2533 (1966).
9. Monod, J., J.-P. Changeux, and F. Jacob, *J. Mol. Biol.,* **6,** 306 (1963).
10. Larsson, A., and P. Reichard, *J. Biol. Chem.,* **241,** 2540 (1966).
11. Brown, N. C., A. Larsson, and P. Reichard, *J. Biol. Chem.,* **242,** 4272 (1967).
12. Brown, N. C., R. Eliasson, P. Reichard, and L. Thelander, unpublished results.
13. Canellakis, Z. N., and P. Reichard, unpublished observations.
14. Crawford, I. P., and C. Yanofsky, *Proc. Natl. Acad. Sci. U. S.,* **44,** 1161 (1958).
15. Gerhart, J. C., and H. K. Schachman, *Biochemistry,* **4,** 1054 (1965).
16. Hummel, J. P., and K. J. Dreyer, *Biochim. Biophys. Acta,* **63,** 530 (1962).
17. Brown, N. C., R. Eliasson, P. Reichard, and L. Thelander, *Biochem. Biophys. Res. Commun.,* **30,** 522 (1968).

18. Changeux, J.-P., J. C. Gerhart, and H. K. Schachman, *Biochemistry,* **7,** 531 (1968).
19. Vagelos, P. R., A. W. Alberts, and D. B. Martin, *J. Biol. Chem.,* **238,** 533 (1963).
20. Datta, P., H. Gest, and H. L. Segal, *Proc. Natl. Acad. Sci. U.S.,* **51,** 125 (1964).
21. Phillips, A. T., and W. A. Wood, *Biochem. Biophys. Res. Commun.,* **15,** 530 (1964).
22. Hirata, M., M. Tokushige, A. Inagaki, and O. Hayaishi, *J. Biol. Chem.,* **240,** 1711 (1965).
23. Iwatsuki, N., and R. Okazaki, *J. Mol. Biol.,* **29,** 139 (1967).
24. Moore, E. C., and R. B. Hurlbert, *J. Biol. Chem.,* **241,** 4802 (1966).
25. Lark, K. G., in *Molecular Genetics,* J. H. Taylor, editor, vol. I, Academic, New York, 1963, p. 153.
26. Klenow, H., *Biochem. Biophys. Acta,* **35,** 412 (1959).
27. Prusoff, W. H., *Biochem. Pharmacol.,* **2,** 221 (1959).
28. Munch-Petersen, A., *Biochem. Biophys. Res. Commun.,* **3,** 392 (1960).
29. Morris, N. R., and G. A. Fischer, *Biochem. Biophys. Acta,* **42,** 183 (1960).
30. Klenow, H., *Biochem. Biophys. Acta,* **61,** 885 (1962).
31. Morris, N. R., and G. A. Fischer, *Biochem. Biophys. Acta,* **68,** 84 (1963).
32. Gentry, G. A., P. A. Morse, Jr., and V. R. Potter, *Cancer Res.,* **25,** 517 (1965).
33. Whittle, E. D., *Biochem. Biophys. Acta,* **114,** 44 (1966).
34. Odmark, G., and B. A. Kihlman, *Mutation Res.,* **2,** 274 (1965).